SpringerBriefs in Applied Sciences and Technology

Thermal Engineering and Applied Science

Series Editor

Francis A. Kulacki

For further volumes:
http://www.springer.com/series/10305

Matthew Lind Roesle · Francis A. Kulacki

Boiling Heat Transfer in Dilute Emulsions

 Springer

Matthew Lind Roesle
Institute of Energy Technology
ETH Zurich
Zurich
Zürich
Switzerland

Francis A. Kulacki
Department of Mechanical Engineering
University of Minnesota
Minneapolis, MN
USA

ISSN 2193-2530   ISSN 2193-2549   (electronic)
ISBN 978-1-4614-4620-0   ISBN 978-1-4614-4621-7   (eBook)
DOI 10.1007/978-1-4614-4621-7
Springer New York Heidelberg Dordrecht London

Library of Congress Control Number: 2013936970

Printed on acid-free paper

Springer is part of Springer Science+Business Media (www.springer.com)

Preface

Boiling of dilute emulsions in which the dispersed component has a lower boiling point than the continuous component has received little attention in the literature. These mixtures exhibit several surprising behaviors that were unknown until the 1970s. Boiling of the dispersed component enhances heat transfer over a wide range of surface temperatures without transition to film boiling, but a high degree of superheat is required to initiate boiling. In single-phase convection, the dispersed component has little effect on heat transfer. These behaviors appear to occur in part because few droplets in the emulsion contact nucleation sites on the heated surface. No physically consistent model of boiling in dilute emulsions exists at present.

The unusual behavior of boiling dilute emulsions makes them potentially useful for high heat flux cooling of electronics. High-power electronic devices must be maintained at junction temperatures below $\sim$95 °C to operate reliably, even while generating heat fluxes of 100 W/cm^2 or more. Military avionics is pushing the thermal envelop to junction temperatures of 95 °C and power densities of 1,000 W/cm^3. Current research, generally focusing on single phase convection or flow boiling in small diameter channels, has not yet identified an adequate solution. An emulsion of refrigerant in water would be well-suited to this application. The emulsion retains the high specific heat and thermal conductivity of water, while boiling of the refrigerant enhances the heat transfer coefficient at temperatures below the saturation temperature of water.

We summarize the recent experiments on boiling heat transfer from a heated wire in emulsions of pentane in water and FC-72 in water. These emulsions have properties suitable for practical use in high heat flux cooling applications, unlike most emulsions that have previously been studied. Experiments include enhanced boiling of the continuous component, which has not previously been observed, in addition to boiling of the dispersed component. In both boiling regimes the heat transfer coefficient is enhanced compared to that of water. Visual observation of the boiling process reveals the presence of large attached bubbles on the heated wire, the formation of which coincides with the inception of boiling in the heat transfer data. At very low dispersed component fractions and low temperatures, boiling of individual dispersed droplets is not observed. The large attached bubbles represent a new boiling mode that has not been reported and is, under some circumstances, the dominant mode of boiling heat transfer.

A model of boiling dilute emulsions is described and developed based upon the Euler–Euler model of multiphase flows. The general balance equations are applied to the present situation, thus providing a rigorous and physically consistent framework. The model contains three phases that represent the continuous component, liquid droplets of the dispersed component, and bubbles that result from boiling of individual droplets. Mass, momentum, and energy transfer between the phases are based upon the behavior of and interaction between individual elements of the dispersed phases. A one-dimensional simulation of a single boiling droplet in superheated liquid is used to develop the closure equations of the larger model. Droplet boiling is assumed to occur when a droplet contacts a heated surface or a vapor bubble. Collisions between droplets and bubbles and chain boiling of closely spaced droplets are considered. The model is limited to the dispersed component boiling regime and thus does not account for phase change of the continuous component. It also does not account for large attached bubbles revealed in the experiments. Exercising the model produces several trends observed in the experiments.

Zurich, Switzerland   Matthew Lind Roesle
Minneapolis, MN, USA   Francis A. Kulacki

Contents

Symbols

A	Surface area, m^2
A_0	Adjustment parameter in Eq. 2.17
Ar	Archimedes number, $(\rho_{\text{eff}} - \rho_{\text{b}})\text{g}\rho\text{d}\frac{\text{d}_{\text{b}}^3}{\mu_{\text{eff}}^2}$
a	Coefficient in discretized differential equations
B	Pre-exponential factor in Eq. 2.2
b	Body force per unit mass, N/kg
C_{D}	Drag coefficient
C_{L}	Lift coefficient
C_{R}	Rotation coefficient
C_{sf}	Empirical constant, Eq. 5.3
C_{td}	Turbulent drag coefficient
C_{vm}	Virtual mass coefficient
c_{p}	Constant-pressure specific heat, J/kg °C
c_{v}	Constant-volume specific heat, J/kg °C
D	Viscous dissipation, W/m^3
d	Diameter, m
E	Averaged interfacial energy transfer rate, Eq. (A23), W/m^3
e	Specific internal energy, J/kg
F	Source flow strength, m^3/s
$\boldsymbol{F}$	Averaged interfacial force, N/m^3
$\boldsymbol{F}_{ij}$	Averaged interfacial force on phase i by phase j, N/m^3
f	Frequency, Hz
g	Gravitational force, N/kg
h	Heat transfer coefficient, W/m^2 °C
I	Identity matrix
i	Specific enthalpy, J/kg
J	Volumetric nucleation or collision rate, 1/m^3 s
Ja	Jakob number, $\rho_{\text{f}}c_{\text{p,f}}\Delta T/(\rho_{\text{g}}i_{\text{fg}})$

K_T	Pseudo-turbulent factor
K	Thermal conductivity, W/m °C
k_B	Boltzmann constant, 1.38065×10^{-23} J/K
l	Eddy mixing length
L	Characteristic length, m
m	Mass, kg
$\dot{m}$	Volumetric mass transfer rate from droplet phase to bubble phase, kg/m^{3}s
N	Number density, 1/m^3
$\boldsymbol{n}$	Outward-pointing unit normal vector
Nu	Nusselt number, hL/k
P	Pressure, N/m^2
Pe	Peclet number, Re · Pr
Pr	Prandtl number, v/α
Q	Heat transfer, J
$\boldsymbol{q}$	Heat flux vector, W/m^2
q	Heat transfer rate, W
q''	Heat flux, W/m^2
$\boldsymbol{R}_{\text{eff}}$	Effective viscous stress (molecular plus Reynolds), m^2/s^2
$\mathbf{R}_{\text{eff}}^{\text{C}}$	Correction stress component, m^2/s^2
$\mathbf{R}_{\text{eff}}^{\text{D}}$	Diffusive stress component, m^2/s^2
R	Droplet or bubble radius, m
R_G	Gas constant, J/kg K
R_{cr}	Critical bubble radius, m, Eq. 2.1
Ra	Rayleigh number, $\frac{g\beta_{\text{film}}(T_s - T_\infty)d^3}{v_{\text{film}}^2}\text{Pr}_{\text{film}}$
r	Radial distance, m
Re	Reynolds number, $\mathscr{U}d/v$
$\boldsymbol{S}$	Surface area vector, m^2
S	Heat source per unit mass, W/kg
s	Prandtl number factor, Eq. 5.3
St	Stefan number, $c_{\text{p,d}}(T - T_{\text{sat}})/i_{\text{fg}}$
T	Temperature, K
ΔT	Temperature difference, $T_\infty - T_{\text{sat}}$
$\mathbf{T}$	Stress tensor, N/m^2
t	Time, s
Δt	Time step size in numerical solver, s
$\mathscr{U}$	Characteristic fluid velocity, m/s
U	Velocity vector, m/s
$U_{\text{r},ij}$	Relative velocity vector between phases i and j, $U_j - U_i$, m/s
u	Velocity component, m/s
u^*	Turbulence characteristic velocity, m/s
V	Volume, m^3

v	Specific volume, m^3/kg
X	Phase indicator function
x	Position vector

Greek Symbols

α	Thermal diffusivity, m^2/s
β	Volumetric expansion coefficient, m^3/m^3 K
Γ	Averaged interfacial mass transfer rate, Eq. (A24), kg/m^3 s
γ	Generic diffusive flux, Eq. (A1)
γ	Specific heat ratio, c_p/c_v
Δ	Very small value
δ	Dirac delta function
δ_t	Thermal boundary layer thickness, m
ε	Volume fraction, m^3/m^3
ζ	Generic source density, Eq. (A1)
η	Collision efficiency
θ	Angular direction, radians
κ	Polytropic coefficient
μ	Dynamic viscosity, kg/m-s
v	Kinematic viscosity, m^2/s
ρ	Density, kg/m^3
σ	Surface tension, N/m
τ	Characteristic time, s
ϕ	Volumetric flux, m^3/s
φ	Number of boiling droplets in a chain reaction
Ψ	Generic conserved quantity, Eq. (A1)

Subscripts

$+$	Positive portion
$-$	Negative portion
0	Reference
1ph	Single phase
b	Vapor bubble phase (emulsified component)
c	Continuous phase
coll	Collisions
cond	Condensation
D	Drag
d	Liquid droplet phase (emulsified component) or dispersed phase
eff	Effective value for the emulsion
F	Face-centered value

f	Saturated liquid
fg	Difference between saturated vapor and saturated liquid
film	Evaluated at film temperature, $(T_s + T_\infty)/2$
g	Saturated vapor
I	Interface
inertial	Inertial
init	Initial
i, j	Counting indices, i, j = 1, 2, 3...
L	Lift
M	Minnaert
m	Mixture
max	Maximum value
mol	Molecule
R	Rotational
r	Radial direction
s	Surface
sat	Saturated condition
T	Turbulent quantity
td	Turbulent dispersion
v	Vapor phase
vm	Virtual mass
w	Wall
wire	Heated wire
θ	Tangential direction (in polar coordinates)
∞	Ambient

Superscripts and Other Notation

$\overline{X}$	Average value or phase property
x'	Fluctuating component of x, where $x = x' + \overline{X}$
$\widehat{X}$	Unit vector
$|\mathbf{X}|$	Magnitude of x
δx	Uncertainty in x
x^*	Predicted values of x in PISO algorithm
x^o	Quantity x at previous iteration
x^{Re}	Fluctuation (Reynolds) quantity
$\dfrac{D_i}{D_t}$	Material derivative for phase i, $\frac{\partial}{\partial_t} + \mathbf{U}_i \cdot \nabla$
$\|\mathscr{L}[x]\|$	Expression arising from implicit discretization of operator $\mathscr{L}$ in terms of x
$\langle X \rangle$	Average of x over adjacent cells
$\mathscr{A}$	Discretized linear system of equations

$\mathscr{A} := \{\ldots\}$ Assignment of the discretized form of the system of linear equations in brackets { ... } to $\mathscr{A}$

$\mathscr{A}_D$ Diagonal matrix coefficients of discretized linear system of equations $\mathscr{A}$

$\mathscr{A}_N$ Off-diagonal matrix coefficients of discretized linear system of equations $\mathscr{A}$

$\mathscr{A}_S$ Source vector of discretized linear system of equations $\mathscr{A}$

$\mathscr{A}_H$ H operator, Eq. 4.29

Chapter 1
Introduction

Keywords Emulsion · Dilute emulsion · Boiling · Convection · Electronics cooling · High heat flux cooling

1.1 Introduction

Boiling has long been recognized as an important heat transfer mechanism and boiling of pure liquids is now fairly well understood, but surprising behavior is encountered in some circumstances. One such area is boiling of dilute emulsions in which the dispersed component has a lower boiling point than the continuous component. Under these conditions, the degree of superheat required for boiling is much higher than for single component liquids, and burnout, i.e., the transition to film boiling, does not occur. This anomalous behavior was discovered in the 1970s (Mori et al. 1978) and has since been studied extensively by Bulanov and co-workers (Bulanov and Gasanov 2007, 2008). Despite the experimental studies that have been performed, there is as yet no detailed understanding of how the boiling process occurs.

An emulsion is a mixture of two immiscible liquids in which the dispersed or droplet component forms a suspension of many small droplets in the continuous component. An emulsion is considered dilute when the dispersed component occupies ~5 % or less of the emulsion by volume. In the emulsions considered in this monograph, the dispersed component has a lower boiling point than the continuous component. We focus on heat transfer mechanisms and characteristics of boiling when the only the dispersed component boils. We also describe recent experiments when surface temperatures are large enough so that both components boil.

Superheated droplets in an emulsion exist in a meta-stable state, meaning that they remain liquid despite having a temperature well above their saturation temperature. The meta-stable liquid would rapidly boil, but it must either first contact a liquid–vapor interface or experience some sort of disturbance that initiates boiling. Pure liquids are not often found in meta-stable states because the walls of

M. L. Roesle and F. A. Kulacki, *Boiling Heat Transfer in Dilute Emulsions*, SpringerBriefs in Thermal Engineering and Applied Science, DOI: 10.1007/978-1-4614-4621-7_1, © The Author(s) 2013

the container holding them contain nucleation sites, microscopic cavities resulting from the manufacturing process that retain gases. The theoretical limit of the degree of superheat is a function of the liquid and the ambient pressure, but is typically 100 °C or more and is also greater than the point at which boiling occurs in dilute emulsions (Bulanov and Gasanov 2007, 2008). An understanding of the mechanisms that cause individual droplets in the emulsion to boil is crucial to understanding and predicting the overall behavior of boiling emulsions.

Bulanov and Gasanov (2008) give a possible explanation for droplets boiling in emulsions at temperatures below the theoretical limit of superheat based on chain-reaction boiling of the droplets due to the presence of impurities in the liquid. The droplets of the low-boiling-point liquid contain floccules of nanoparticles, too small to be detected or filtered out by normal means. These floccules contain some trapped atmospheric gases absorbed on their surfaces and can act as nucleation sites for boiling. They speculate that the rapid expansion of a boiling droplet creates a shockwave that breaks up floccules in any nearby droplets, thus causing those droplets to immediately boil as well. While this explanation agrees with much of their experimental data in a qualitative sense, the physical processes are not modeled in detail, and there are insufficient details in their papers to allow prediction of the behavior of boiling emulsions. Some experimental data for emulsions in which particles have been added to the low-boiling-point liquid are contradictory as well (Bulanov et al. 2006).

The unusual characteristics of boiling dilute emulsions may be useful in addressing an open problem in heat transfer engineering: high heat flux cooling of electronics. High-power electronic processors must be maintained at junction temperatures below ~ 85 to 95 °C to operate reliably, even while producing heat fluxes as high as 100 W/cm^2 (Thome 2006) and power densities up to 1,000 W/cm^3 in military avionics. The bulk of recent research to address these requirements have focused on single phase and boiling heat transfer in small diameter channels. Often called microchannels, these channels can be etched directly into the silicon substrates of the electronic device, thus providing an extremely short heat conduction path for the heat generated in the electronic junctions (Tuckerman and Pease 1981).

Experiments performed by Tuckerman and Pease demonstrate that high heat flux cooling can be achieved using single-phase convection to water in this manner. However, at high heat flux, the temperature gradient along the length of the microchannel is large, and high inlet pressure is required to force water through the microchannels at a sufficient rate. Both of these effects place mechanical stress on the silicon substrate and lower the performance for the overall thermal management system. For single-phase convection in simple microchannels, tradeoffs between small temperature rise in the liquid, low pressure drop, and low temperature difference between surface and fluid are unavoidable. Various methods of improving on this situation have been tried including placing structures in the microchannel (Kandlikar and Grande 2004), electrical fields, and vibrating elements (Steinke and Kandlikar 2004). The effect of these changes is to break up the laminar flow structure of the liquid in the microchannel, which improves heat transfer at the cost of increased pressure drop and lower coefficient of performance.

A separate set of problems is associated with the use of boiling heat transfer for high heat flux cooling. The most significant problem is that the heat transfer fluid of choice, water, has too high a saturation temperature to be used for cooling electronics. Other refrigerants and heat transfer liquids have lower saturation temperatures, but they also have much lower critical heat flux (CHF), typically well below 100 W/cm^2 (Wojtan et al. 2006; Zhang et al. 2007; Agostini et al. 2007). Similar to single-phase heat transfer, some improvement in CHF can obtained through the use of more complex geometries, such as arrays of micro-jets directed into the microchannels (Sung and Mudawar 2009). However, changes in geometry alone produce only modest improvements in CHF, and more significant improvements have been demonstrated only by cooling the refrigerant to well below room temperature and allowing the surface temperature to exceed 85 °C as well (Sung and Mudawar 2009).

Boiling in a dilute emulsion offers an alternative to single-phase and boiling heat transfer that combines the positive characteristics of both heat transfer mechanisms. In a dilute emulsion of refrigerant in water, water makes up the bulk of the emulsion and gives the mixture large heat capacity and high thermal conductivity. With the proper choice of the dispersed component, the droplets will boil rapidly at temperatures below the saturation temperature of the continuous fluid, thus agitating the water flow and breaking up its laminar flow structure (Bulanov et al. 1996). Such agitation of the water may improve heat transfer rates far more effectively than placing structures in the channel, and will do so without increasing the complexity of the channel itself. An experimental study in which water and a refrigerant are introduced into a flat microchannel in parallel streams highlights the importance of the intimate contact between refrigerant and water that is achieved in emulsions. When so separated, any boiling that occurs in the refrigerant has negligible effect on the water stream, and no heat transfer enhancement is obtained over the case of a water-only flow (Roesle and Kulacki 2008).

At present, fundamental understanding of boiling emulsions is very limited. The most detailed description of how emulsions boil is that of Bulanov and Gasanov (2008) but leaves many questions unanswered. The structure of the boundary layer in a boiling emulsion near a heated surface is not known. It is not known precisely where or how boiling occurs within the boundary layer, and it is not known how the boiling droplets and bubbles influence each other. Also, the model is not predictive of the heat transfer rate for any given set of conditions or surface.

In this monograph, we review the status of the field through 2012, describe the explosive boiling of a single droplet in carrier component of higher saturation temperature, and develop and evaluate a physical model of dilute emulsions undergoing boiling. Our model for the dilute emulsion is rooted in a detailed analysis of the behavior of a single highly superheated droplet undergoing boiling. The effects of the boiling droplet on the surrounding fluid and therefore any nearby droplets are also considered. These boiling phenomena are linked to the overall behavior of the emulsion using the Euler–Euler approach to modeling multiphase mixtures (Rusche 2002). Experiments examine water-FC-72 and water-pentane emulsions undergoing free convection boiling on a horizontal heated wire and offer an opportunity to observe the detailed behavior of the boiling process. Visual observation of the heat transfer surface during boiling correlated with the heat transfer data are a new contribution to the literature.

Given the early stage of development of this field, our approach in this monograph is to identify all the relevant physical processes and mechanisms in boiling dilute emulsions and to build a model up from basic principles that incorporates them. Such a physically complete and consistent model is necessary before meaningful work can be done on refining components of the model or incorporating advanced modeling techniques. The model developed here goes considerably towards this goal, although the reader will see how the current focus on the behavior of the bulk of the emulsion leaves shortcomings in the handling of interactions between the dispersed component of the emulsion and solid surfaces. Nevertheless, we hope that the model and experimental findings described here will spark further research in this field and will form the foundation that allows more refined and advanced modeling of boiling heat transfer in dilute emulsions.

References

Agostini B, Fabbri M, Park JE et al (2007) State of the art of high heat flux cooling technologies. Heat Transf Eng 28:258–281

Bulanov NV, Gasanov BM (2007) Special features of boiling of emulsions with a low-boiling dispersed phase. Heat Transf Res 38:259–273

Bulanov NV, Gasanov BM (2008) Peculiarities of boiling of emulsions with a low-boiling disperse phase. Int J Heat Mass Transf 51:1628–1632

Bulanov NV, Skripov VP, Gasanov BM, Baidakov VG (1996) Boiling of emulsions with a low-boiling dispersed phase. Heat Transf Res 27:312–315

Bulanov NV, Gasanov BM, Turchaninova EA (2006) Results of experimental investigation of heat transfer with emulsions with low-boiling disperse phase. High Temp 44:267–282

Kandlikar SG, Grande WJ (2004) Evaluation of single phase flow in microchannels for high heat flux chip cooling—thermohydraulic performance enhancement and fabrication technology. Heat Transf Eng 25:5–16

Mori YH, Inui E, Komotori K (1978) Pool boiling heat transfer to emulsions. J Heat Transf Trans ASME 100:613–617

Roesle ML, Kulacki FA (2008) Characteristics of two-component two-phase flow and heat transfer in a flat microchannel. In: Proceedings of the 2008 ASME summer heat transfer conference. ASME, New York

Rusche H (2002) Computation fluid dynamics of dispersed two-phase flows at high phase fractions. Ph.D. Thesis, University of London, London

Steinke ME, Kandlikar SG (2004) Single-phase heat transfer enhancement techniques in microchannel and minichannel flows. In: Proceedings of the second international conference on microchannels and minichannels. ASME, New York

Sung MK, Mudawar I (2009) CHF determination for high-heat flux phase change cooling system incorporating both micro-channel flow and jet impingement. Int J Heat Mass Transf 52:610–619

Thome JR (2006) The new frontier in heat transfer: Microscale and nanoscale technologies. Heat Transf Eng 27:1–3

Tuckerman DB, Pease RFW (1981) High-performance heat sinking for VLSI. IEEE Electron Device Lett 2:126–129

Wojtan L, Revellin R, Thome JR (2006) Investigation of saturated critical heat flux in a single, uniformly heated microchannel. Exp Therm Fluid Sci 30:765–774

Zhang H, Mudawar I, Hasan MH (2007) Assessment of dimensionless CHF correlations for subcooled flow boiling in microgravity and Earth gravity. Int J Heat Mass Transf 50:4568–4580

Chapter 2
Status of the Field

Keywords Meta-stable liquid · Spontaneous nucleation · Superheat limit · Multiphase flow · Eulerian multiphase model · Emulsion · Boiling heat transfer · Chain boiling

The literature on boiling of emulsions is relatively sparse, and very little theoretical work on the subject exists. In this chapter, the literature on boiling emulsions is reviewed, as well as some related topics that are important to understanding the main subject. These topics include spontaneous nucleation in liquids and modeling of multiphase droplet and particulate flows. Because boiling in dilute emulsions entails some physical processes unique in multiphase flow, the discussion of multiphase flow modeling begins from fundamentals and is generally limited to basic and generic dispersed multiphase modeling techniques. Sophisticated models are not necessary for our primary goal of identifying the relevant physical processes and significant terms in the balance equations of multiphase flow models, to be undertaken in Chap. 4. We also feel that such techniques should be incorporated only after carefully determining that they are applicable to the problem of boiling in dilute emulsions.

2.1 Meta-Stable Liquid and Spontaneous Nucleation

Boiling in liquids always begins with the nucleation of vapor bubbles. For a heated surface placed in a pure liquid, bubble nucleation occurs at the surface. The microscopic pits and grooves found on most solid surfaces provide numerous nucleation sites, and so nucleate boiling usually begins when the temperature of the heated surface is only a few degrees above the saturation temperature of the liquid. In contrast, dilute emulsions usually do not begin boiling until the heated surface temperature is several tens of degrees higher than the saturation temperature of the dispersed component. The droplets in the emulsion may become

M. L. Roesle and F. A. Kulacki, *Boiling Heat Transfer in Dilute Emulsions,*
SpringerBriefs in Thermal Engineering and Applied Science,
DOI: 10.1007/978-1-4614-4621-7_2, © The Author(s) 2013

Fig. 2.1 Isotherm of a Van der Waals fluid

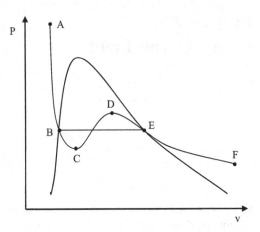

superheated by passing through the thermal boundary layer on a heated surface without contacting it and its nucleation sites. Thus, it is worthwhile to review nucleation in superheated liquid and the limits of liquid superheat.

Curve ABCDEF in Fig. 2.1 represents an isotherm for a fluid obeying the Van der Waals or similar equation of state with temperature less than its critical temperature. If the fluid slowly expands isothermally from a compressed liquid state (state A), it will proceed along the isotherm to state B where it is saturated liquid. The expected behavior as the liquid expands further is for nucleation sites to become active on the walls of the vessel containing the fluid. The fluid will then become a saturated mixture and follow line BE. If no nucleation sites are present, the liquid will instead continue along the isotherm to state C as meta-stable liquid. This process continues only up to state C. It is impossible for a Van der Waals fluid to exist in stable equilibrium between states C and D because $\partial P / \partial V > 0$. State C is thus the thermodynamic limit of superheat for the fluid (Blander and Katz 1975). When a liquid reaches state C it will boil without the presence of any nucleation sites or any external disturbances, a process known as spontaneous nucleation.

The segment CD of the isotherm is known as the Van der Waals loop, and any fluid that exhibits such a loop will have a meta-stable liquid state with an upper bound where the slope of the P–v curve crosses zero. Measurements have shown that the Van der Waals equation of state under predicts the superheat limit for most fluids (Blander and Katz 1975). Other equations of state give more accurate results, although there is some uncertainty associated with extrapolating the state of meta-stable fluids using data from the stable liquid and vapor regions.

The limit of superheat can also be predicted using kinetics. Density fluctuations are always present in liquids, which form microscopic vapor bubbles, or bubble embryos. Most bubble embryos are ephemeral and immediately collapse on themselves due to surface tension. In a highly superheated liquid, some bubbles may pass a critical radius,

$$R_{cr} = \frac{2\sigma}{P - P_\infty}, \qquad (2.1)$$

and continue to grow. At the critical radius the internal pressure, P, exactly balances the ambient pressure and surface tension, but this mechanical equilibrium is unstable. An embryo that is larger than the critical radius will grow to macroscopic size, while an embryo smaller than the critical radius will collapse. Assuming the formation of the bubble embryo occurs isothermally, the pressure in the bubble is equal to the vapor pressure of the liquid at the ambient temperature, $P_v(T_\infty)$. The rate at which bubble embryos grow to the critical radius depends on the reaction kinetics, where the reactions under consideration are evaporation from and condensation to the bubble wall. The kinetic limit of superheat is defined as the temperature at which the rate of bubble nucleation due to density fluctuations becomes large. If the bubble embryo remains in mechanical equilibrium as it grows, the nucleation rate in a superheated liquid is,

$$J = N_{mol} \left(\frac{2\sigma}{\pi m_{mol} B} \right)^{1/2} \exp\left[\frac{-16\pi\sigma^3}{3k_B T (P_v - P_\infty)^2} \right] \qquad (2.2)$$

where $B = 2/3$ (Blander and Katz 1975).

The assumption of mechanical equilibrium may not be correct under all circumstances. Bubble embryo growth can be limited by four alternative factors: viscosity, inertia, rate of evaporation, and heat transfer. In each of these cases the pre-exponential factor of J differs, but the exponent remains unchanged (Kagan 1960). For example, assuming that the embryo remains in chemical equilibrium $(P = P_v)$ instead of mechanical equilibrium $(P = P_\infty + 2\sigma/R)$, the value of B changes from 2/3 to 1. Owing to the rapid rise in the exponential term with temperature, such minor changes in the pre-exponential factor are irrelevant for most liquids (Blander and Katz 1975). The effects of non-ideal gas behavior in the bubble embryo are similarly negligible when considering spontaneous nucleation of a pure substance (Katz and Blander 1973).

Unlike the thermodynamic limit of superheat, this description of spontaneous nucleation does not provide a definite temperature at which spontaneous nucleation will occur. In fact, Eq. (2.2) predicts a nonzero rate of nucleation for any superheated liquid. To define the kinetic limit of superheat from this theory, one must also specify a nucleation rate above which the liquid is considered to be boiling. In practice, values between 10^{12} and 10^{20} m^{-3} s^{-1} are generally found to give good agreement with experiment (Blander and Katz 1975; Chen et al. 2006). This range is large, but the exponential term of Eq. (2.2) changes extremely rapidly with temperature for pure liquids and so corresponds to a temperature range of less than $3°$. As shown in Table 2.1 for Fluorinert[TM] FC-72, J increases by a factor of approximately 10^3/K near its limit of superheat. Other pure liquids have similar rates of increase in nucleation rate.

Spontaneous nucleation may be produced in the laboratory either by eliminating nucleation sites or by heating a surface rapidly enough that nucleate boiling at the

Table 2.1 Nucleation rates of Fluorinert™ FC-72 at 1 atm (Chen et al. 2006)

T (K)	P_{sat} (kPa)	ρ_f (kg/m³)	σ (N/m)	J (m^{-3} s^{-1})
403.2	729.1	1320.9	0.0026	5.26×10^{-24}
404.2	745.0	1311.7	0.0025	1.24×10^{-14}
405.2	761.1	1302.8	0.0025	3.97×10^{-8}
406.2	777.5	1293.9	0.0024	2.26×10^{-2}
407.2	794.1	1284.9	0.0023	2.78×10^{3}
408.2	811.0	1276	0.0023	8.71×10^{7}
409.2	828.2	1267.1	0.0022	0.81×10^{10}
410.2	845.7	1258.1	0.0021	4.40×10^{12}
411.2	863.4	1249.2	0.0021	2.55×10^{15}
412.2	881.3	1240.3	0.0020	3.07×10^{18}
413.2	899.6	1232.8	0.0020	1.57×10^{21}

surface is not significant. The earliest studies of spontaneous nucleation used drawn and annealed glass capillary tubes, which have exceptionally smooth surfaces with few or no nucleation sites (Wismer 1922). The pulsed heating method uses electrical heat to heat a surface at a rate of 10^7 K/s or more, which is fast enough that the superheat limit is reached and spontaneous nucleation can be observed before the surface becomes covered with bubbles formed by nucleate boiling (Iida et al. 1994; Pavlov and Skripov 1970). The floating droplet method uses a column of dense liquid that is heated unevenly. A droplet of a less dense liquid is introduced at the bottom, cool end of the column and rises into ever-warmer liquid until it boils by spontaneous nucleation (Wakeshima and Takata 1958). This method has been used for a wide variety of liquids and experimental results generally agree well with the kinetic limit of superheat (Blander and Katz 1975, Skripov 1974). These studies are usually carried out with highly purified liquids. It has been shown that solid particles suspended in liquid do not provide nucleation sites if all gas entrapped in them has been driven off (Buivid and Sussman 1978).

2.2 Continuum Models of Multiphase Flow

Multiphase flows[1] exist in a variety of forms in nature and in many industrial processes. Some examples include the flow of sediment-laden river water, boiling liquids, sprays of liquid fuel in combustors, and the emulsions that are the subject

---

[1] The terms *phase* and *component* should be clarified, as their usage is not uniform across all areas of multiphase flow. When discussing emulsions in this monograph, component refers to a distinct substance, while phase identifies both substance and state of matter. For example, a boiling emulsion of FC-72 in water contains two components (FC-72 and water) but three phases (vapor FC-72, liquid FC-72, and liquid water). In contrast, in the multiphase flow modeling literature, phase is often used generically and could refer to either a component or a phase, depending on how the model is applied. In this chapter, phase is used in this generic sense.

of this study. These examples are classified as dispersed flows, because in each case one phase exists as many separate elements distributed throughout the other phase. Although the Navier–Stokes equations that govern the motion of these flows are well known, it is generally impossible, or at least impractical, to solve them directly for the flow field in dispersed multiphase flows because there are too many surfaces to track individually.

Given the complexity of multiphase flows, any practical flow model must account for the interaction between phases without tracking all the interfaces between them. Doing so necessarily requires assumptions regarding the structure of the dispersed phase(s), and these assumptions limit the range of applicability of any model. Consequently many models exist for various types of flows and geometries and at differing levels of complexity, fidelity, and generality (Wallis 1969).

Increasingly, modeling efforts in multiphase flow are directed toward the field of computational fluid dynamics (CFD). Approaches to CFD can generally be divided into three approaches: direct numerical simulation (DNS), Euler–Lagrange, and Euler–Euler (Rusche 2002). These are illustrated schematically in Fig. 2.2. The DNS approach is the most computationally costly, as it requires simulating the (usually very small-scale) motion of the fluids and the surfaces dividing the phases. This approach is often prohibitively costly for flows of engineering interest. In the Euler–Lagrange approach, the motions of individual elements of the dispersed phase are simulated but the internal behavior of each element is not. The name Euler–Lagrange reflects the fact that the dispersed phase is simulated using a Lagrangian method (Newton's laws of motion), while the continuous phase is simulated with an Eulerian method (the Navier–Stokes equations). Finally, the Euler–Euler approach is the least computationally intensive because it dispenses with simulating individual elements of the dispersed phase entirely. Instead each phase is represented as a continuous fluid that occupies a fraction, ε, of the total volume (Bouré and Delhaye 1982; Gidaspow 1994).

The straightforward interpretation of the Euler–Euler approach is that all the quantities that are considered (phase fraction, velocity, temperature, and so on) are composite time- and space-averaged values. This approach appears well-suited to CFD wherein the flow domain under consideration is split into a number of discrete volumes and the simulation proceeds by discrete time steps. While this interpretation has clear physical meaning, it leads to a number of difficulties that are discussed in detail by Hill (Hill 1998).

A limitation of composite-averaged equations is in the scale of the discrete volumes and time steps. For a volume-averaged quantity of the dispersed phase to be meaningful, the volume over which averaging takes place must be large compared to the average spacing between the elements of the dispersed phase. Similarly for a meaningful time-averaged quantity, the averaging period must be long compared to the transit time of an element of the dispersed phase through the volume. On the other hand, the size of the discrete volumes and time steps must be small compared to the macroscopic flow features that the simulation is meant to capture. It is not always possible to satisfy both requirements simultaneously (Hill 1998).

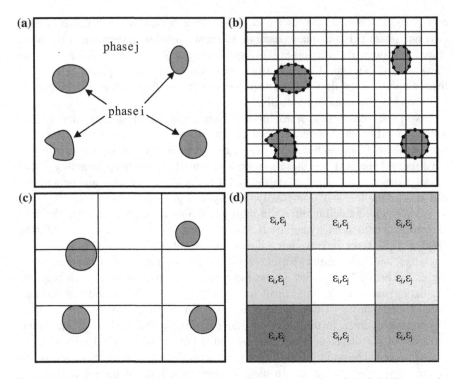

Fig. 2.2 Modeling schemes for two-phase flow. **a** Dispersed two-phase flow. **b** DNS approach where surfaces as well as fluid elements are simulated. **c** Euler–Lagrange approach where individual elements of dispersed phase are simulated, **d** Euler–Euler approach where average volume fraction of each phase is simulated

Another difficulty rises from correlation of fluctuations in the averaged quantities. Usually the composite-averaging process is accomplished by averaging over the discrete volume first and then over time. The intermediate quantity (an instantaneous, space-averaged value) fluctuates in time. A number of new terms arise in the Navier–Stokes equations due to correlations in the fluctuations of different quantities, similar to the Reynolds stress terms found in turbulent flows, and the physical meanings of the new terms are not always clear (Hill 1998).

These difficulties are eliminated by abandoning composite averaging and instead using ensemble averages. This approach, also called conditional averaging, yields averaged mass, momentum, and internal energy equations in a single operation (Drew and Passman 1999),

$$\frac{\partial \varepsilon_i \bar{\rho}_i}{\partial t} + \nabla \cdot (\varepsilon_i \bar{\rho}_i \bar{U}_i) = \Gamma_i \tag{2.3}$$

$$\frac{\partial \varepsilon_i \bar{\rho}_i \overline{U}_i}{\partial t} + \nabla \cdot \left(\varepsilon_i \bar{\rho}_i \overline{U}_i \overline{U}_i\right) = \nabla \cdot \left[\varepsilon_i \left(\overline{T}_i + T_i^{\text{Re}}\right)\right] + \varepsilon_i \bar{\rho}_i \bar{b}_i + F_i + U_{i,I} \Gamma_i \tag{2.4}$$

$$\frac{\partial \varepsilon_i \bar{\rho}_i \bar{e}}{\partial t} + \nabla \cdot (\varepsilon_i \bar{\rho}_i \bar{U}_i \bar{e}_i) = \varepsilon \bar{T}_i : \nabla \bar{U}_i - \nabla \cdot \left[\varepsilon_i (\bar{q}_i + q_i^{Re}) \right]$$
$$+ \varepsilon_i \bar{\rho}_i \bar{S}_i + \varepsilon_i D_i + E_i + e_{i,I} \Gamma_i \tag{2.5}$$

where all quantities in Eqs. (2.3–2.5) are averaged. The over bar denoting averaging is omitted henceforth for brevity. Internal energy, e_i, and velocity, U_i, are mass-weighted (Favré averaged). The body force is b_i and the volumetric heating rate is S_i. The derivation of these equations is described in Appendix A.

The volumetric rate of mass transfer to phase i through interfaces is Γ_i. The interfacial force applied to phase i from other phases is F_i and the interfacial internal energy transfer to phase i is E_i. Stress is decomposed into an average stress, T_i, and the Reynolds stress, T_i^{Re}. Similarly, the heat flux is decomposed into the flux based on average temperature gradients, q_i, and the fluctuation heat flux, q_i^{Re}. The internal energy equation also contains dissipation, D_i. All of these terms must be defined through either constitutive laws or other closure equations.

The fluid is generally assumed to be Newtonian so that under the Stokes' assumption,

$$\mathbf{T} = -P\mathbf{I} + \mu \left(\nabla \mathbf{U} + \nabla^T \mathbf{U} - \frac{2}{3} \mathbf{I} (\nabla \cdot \mathbf{U}) \right). \tag{2.6}$$

It is further assumed that the pressure is the same in each phase. If the Reynolds stress is represented using the Boussinesq eddy viscosity hypothesis (Pope 2000), the stress term in Eq. (2.4) is expressed,

$$\nabla \cdot \left[\varepsilon_i (\mathbf{T}_i + \mathbf{T}_i^{Re}) \right] = -\varepsilon_i \nabla P + \nabla \cdot \left[\varepsilon_i (\bar{\mu}_i + \mu_{i,T}) \left(\nabla \mathbf{U}_i + \nabla^T \mathbf{U}_i - \frac{2}{3} \mathbf{I} (\nabla \cdot \mathbf{U}_i) \right) \right]$$
$$= -\varepsilon_i \nabla P - \nabla \cdot (\varepsilon_i \rho_i \mathbf{R}_{\text{eff},i}) \tag{2.7}$$

For brevity, the effective viscous stress term can be expressed as $\mathbf{R}_{\text{eff},i}$, as shown above. The term $P\nabla \varepsilon_i$ is absorbed into a fluctuating interfacial pressure term by Hill (1998). Hill (1998) and Rusche (2002) use a k-epsilon turbulence closure model and include an additional term that is a function of the turbulent kinetic energy. On the other hand, Drew and Passman (1999) omit the $\nabla \cdot U_i$ term and use different average velocities for the viscous and turbulent stresses.

The interfacial momentum transfer term in Eq. (2.4) is decomposed into several individual forces. Drew and Passman (1999) identify drag, virtual mass, lift, rotation, turbulent dispersion, and other forces. The expressions for these forces are typically based on the forces acting on a single representative element of the dispersed phase, and a correction is sometimes applied to account for interactions between the many dispersed elements present in a multiphase flow.

For flows in which the dispersed elements are small enough to be subjected to Stokes' drag, the averaged drag force is (Ishii and Zuber 1979),

$$\mathbf{F}_D = \frac{18\varepsilon_d}{d_d^2}\mu_{\text{eff}}(\mathbf{U}_c - \mathbf{U}_d), \tag{2.8}$$

where the subscripts d and c refer to the dispersed and continuous phases respectively, d_d is the characteristic diameter of the dispersed elements, and μ_{eff} is the effective viscosity of the mixture. The effective viscosity accounts for the increased drag experienced by dispersed elements when they are not widely spaced (that is, when ε_d is not near zero). Rusche and Issa (2000) develop correction factors as functions of ε_d that may be applied to the single-element drag force for bubbly, droplet, and particulate flows.

Drew and Passman (1999) express virtual mass, lift, and rotational forces as,

$$\mathbf{F}_{\text{vm}} = C_{\text{vm}}\varepsilon_d\rho_c\left(\frac{D_c\mathbf{U}_c}{Dt} - \frac{D_d\mathbf{U}_d}{Dt}\right), \tag{2.9}$$

$$\mathbf{F}_L = -C_L\varepsilon_d\rho_c(\mathbf{U}_c - \mathbf{U}_d) \times (\nabla \times \mathbf{U}_c), \tag{2.10}$$

$$\mathbf{F}_R = -C_R\varepsilon_d\rho_c(\mathbf{U}_c - \mathbf{U}_d) \times (\nabla \times \mathbf{U}_d), \tag{2.11}$$

where $C_{\text{vm}} = \frac{1}{2}$ and $C_L = C_R = \frac{1}{4}$. They note that the principle of objectivity requires that $C_L + C_R = C_{\text{vm}}$, for only in that case is the sum $\mathbf{F}_{\text{vm}} + \mathbf{F}_L + \mathbf{F}_R$ frame indifferent. The three forces, taken individually, are not frame indifferent, but many practitioners consider only the virtual mass force, or virtual mass and lift forces, and neglect the others (Behzadi et al. 2004; Gosman et al. 1992; Hill 1998; Hao and Tao 2003b). Measurements show that for dispersed mixtures, the virtual mass coefficient increases slowly with ε_d (Drew and Passman 1999; Gosman et al. 1992) while the lift coefficient rapidly decreases toward zero with increasing dispersed phase fraction (Behzadi et al. 2004). This discrepancy may be an indication that the form of Eqs. (2.10) and (2.11) is not correct for dispersed flow. Rusche (2002) examines other equations for lift that do not depend on shear in the average flow.

Gosman et al. (1992) and Drew and Passman (1999) include an additional turbulent drag force, which Drew and Passman express as,

$$\mathbf{F}_{\text{td}} = C_{\text{td}}\frac{3}{4}\frac{\rho_c C_d}{d_d}\varepsilon_d|\mathbf{U}_c - \mathbf{U}_d|\mathbf{T}_c \cdot \nabla\varepsilon_d, \tag{2.12}$$

where for small particles $C_{\text{td}} = 1$. Other interfacial forces include the Basset and Faxén forces, as well as a half-dozen additional terms that Drew and Passman (1999) list but do not name. These forces are neglected in simulations of multiphase flow. Gosman et al. (1992) find that only the drag force is significant for liquid–solid flows and that drag and virtual mass forces are significant for liquid–gas flows.

Many practitioners are concerned with describing the flow field for non-reacting two-phase flows, so that $\Gamma_i = 0$, and the internal energy equation is not solved. Hao and Tao (2003a, b) have used the Euler–Euler approach to successfully simulate two-phase flows with both heat and mass transfer between phases.

Hao and Tao numerically predict the melting of a packed bed of ice spheres in liquid channel flow (2003b) and compare their results to experiments (2003a). Additional constitutive relations are required to model heat transfer between the phases due to convection and mass transfer due to melting. They use an energy equation similar to Eq. (2.5) but expressed in terms of enthalpy rather than internal energy. They neglect the viscous dissipation and flow work terms, and use Fourier's law to express $\mathbf{q}_i$ in terms of the mean temperature gradient. Their simulations are for laminar flow so they do not include the turbulent heat flux term. Results of the simulations match experiments well for changes in the depth of the packed bed as well as the mass of melted ice over time.

In a separate study, Hao and Tao (2004) use an Euler–Euler model to simulate flow and heat transfer in a laminar flow of a liquid laden with particles of encapsulated phase change material (PCM). The governing equations are similar to those obtained in the previous study, although they are simpler in this case because there is no mass transfer between the encapsulated PCM and the carrier fluid. The specific enthalpy of the particle component is used to differentiate between solid, partially melted, and liquid PCM, and the bulk properties (density, specific heat) of the PCM are defined accordingly at each point. Thermal equilibrium within each PCM particle is assumed, although not between the PCM and the carrier fluid. The study simulates flow in a microchannel with a diameter of 122 μm and particle sizes between 0.24 and 10 μm. No experimental data under such conditions exist for direct comparison, but the authors find qualitative agreement between their results and experiments performed at larger scales. Further numerical simulations are made with the goal of finding optimal operating conditions for such particle-laden flows (Xing et al. 2006). These simulations are performed at a larger scale so that their conditions can be matched to experiments. The authors again find reasonably good agreement with experiment, although sensitivity of both experiment and simulation to the flow inlet temperature near the melting temperature of the PCM leads to large uncertainty in the results.

Determination of the properties of each phase can be problematic. In an Euler–Euler model of multiphase flow, a collection of discrete dispersed elements is represented as a fictitious continuous fluid, and the thermophysical properties of the fictitious fluid are not the same as those of the fluid that makes up the actual dispersed elements. The property for which this distinction is most important is viscosity. Many correlations exist for the effective viscosity of multiphase mixtures, but in an Euler–Euler model some method is required to split up the effective viscosity between the phases of the mixture. Xing et al. (2006) take the approach of setting the viscosity of the continuous phase equal to that of the pure fluid and assigning the excess viscosity to the dispersed phase using a weighted average based on volume fraction,[2]

---

[2] An over bar on a fluid property indicates the property assigned to the phase in the Euler–Euler model. So, μ_d is the viscosity of the liquid that makes up the droplets in the emulsion, while $\bar{\mu}_d$ is the viscosity assigned to the phase that represents the droplets.

$$\bar{\mu}_c = \mu_c,$$
$$\bar{\mu}_d = (\mu_{\text{eff}} - \varepsilon_c \mu_c)\varepsilon_d^{-1} \tag{2.13}$$

This linear relation between the phase viscosities is assumed by Xing et al. without justification. They find that this approach when implemented in a numerical simulation gives results that more closely match their experimental data than assigning a fixed value of 0.01 kg/m-s to $\bar{\mu}_d$. A more rigorous approach by Soo (1967) for suspensions of solid particles in gas results in a distribution of viscosity based on mass fractions,

$$\bar{\mu}_c = \left(1 - \frac{\varepsilon_d \rho_d}{\varepsilon_c \rho_c}\right)\mu_{\text{eff}'}$$
$$\bar{\mu}_d = \frac{\varepsilon_d \rho_d}{\varepsilon_c \rho_c}\mu_{\text{eff}}. \tag{2.14}$$

Regardless of the method used to assign viscosity to each phase, a suitable relation must be selected to determine the effective mixture viscosity, μ_{eff}.

2.3 Boiling in Emulsions

Several experimental studies of the heat transfer performance of boiling emulsions have been performed over the past few decades. Emulsions are opaque owing to light scattering at droplet surfaces, and thus direct observation of the boiling process in experiments is generally impossible. Thus, little is known about the detailed behavior of a boiling emulsion at the heated surface and what interactions may occur between the components.

An early study of boiling emulsions by Mori et al. (1978) investigates boiling emulsions of water and oil. They measure heat transfer from a thin heated wire (200 μm diameter, 70 mm length) placed horizontally in a quiescent pool of ~400 ml of emulsion. The experiments are carried out at atmospheric pressure and the bulk temperature is maintained at 100 ± 0.4 °C, so that the water in the emulsion is not significantly subcooled. The oils used have boiling points of at least 196 °C, so the water is the low boiling-point liquid. Their study covers a wide range of mass fractions of water, from 10 to 95 %, and none of the emulsions are therefore considered dilute. They use emulsifiers to produce stable emulsions with average droplet size less than 6 μm.

Mori et al. find that for oil in water emulsions, a significant temperature overshoot occurs before boiling is initiated, as shown in Fig. 2.3a. The designation *oil in water* indicates that water makes up more than half of the mixture, so that the structure of the emulsion may be assumed to be droplets of oil dispersed in water. They suggest that the large temperature overshoot is a result of partial wetting of the heated wire by the oil. Once boiling begins the oil is driven off, causing a subsequent decrease in surface temperature. They also find that, in a few cases of

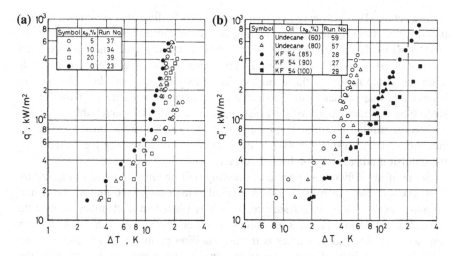

Fig. 2.3 Heat transfer data of Mori et al. for (**a**) oil in water emulsions using KF 96 synthetic oil and 1 % Tween[TM] 80 emulsifier and (**b**) water in oil emulsions using Span[TM] 80 emulsifier. x_o is the mass percent of oil (Mori et al. 1978)

boiling at high heat flux, the surface temperature may suddenly increase by ~ 200 °C, which they speculate may be due to rewetting of the wire by oil. They find that the heat transfer coefficient of the emulsion may be better or worse than that in water depending on the emulsifier, but the type of oil generally does not have any effect. The boiling heat transfer coefficient generally decreases with increasing oil concentration.

Mori et al. also study water in oil emulsions and find that some of the trends identified in oil in water emulsions continue, as seen in Fig. 2.3b. Water in oil emulsions require higher temperature overshoots before boiling begins, typically between 50 and 100 °C, and have lower heat transfer coefficients. Unlike the oil in water cases, the type of oil has an effect on the heat transfer coefficient. Mori et al. do not speculate on the boiling mechanism of water in oil emulsions. In some runs the experiment is halted after a "blowing-up of emulsion due to a sudden foaming in the bulk," which presumably results from the bulk temperature rising slightly above 100 °C, so that when some trigger occurs a large amount of water boils all at once.

A study by Mori et al. (1980) links heat transfer measurements in boiling water in oil emulsions to observations of the heated surface performed using a high-speed video camera. They are able to overcome the opacity of emulsions and observe boiling at a heated surface only under a narrow range of conditions. A thin layer of emulsion with ~ 1 mm depth above the heated surface is used, and the emulsions are prepared with relatively large droplets. Only emulsions with dispersed phase volume fractions between 10 and 20 % produce boiling at moderate heat flux while still leaving boiling on the heated surface visible. At high heat fluxes the number of bubbles becomes large enough to obscure the boiling process.

Mori et al. (1980) observe two forms of boiling: boiling of accumulated droplet liquid on the heated surface, and boiling of droplets in the vicinity of (but not in contact with) the heated surface. Both forms of boiling are generally observed simultaneously in boiling emulsions, and the relative amount of each form depends upon the composition of the emulsion. They hypothesize that the difference in behavior is due primarily to the difference in relative wettability of the heated surface by the components of the emulsions. The degree to which bubbles formed by boiling droplets accumulate on the heated surface also varies between different emulsions. These results emphasize the importance of observing the behavior of the emulsion adjacent to the heated surface.

Some of the difficulties specific to studying boiling emulsions are illustrated by Ostrovskiy (1988). Ostrovskiy uses a vertical cylinder with 6.6 mm diameter and 70 mm length as the heated surface, which is placed in a chamber containing ~ 1.2 L of emulsion. He uses emulsions of water and various other liquids with low boiling points, so that water is the high boiling point liquid. He does not use emulsifiers and instead uses an agitator inside the apparatus to emulsify the mixture and maintain the emulsion. He does not report the droplet sizes of the emulsions. The opacity of the mixture is used as an indicator that the mixture is emulsified.

For emulsions of water and R-113, Ostrovskiy finds that emulsions containing 20 and 40 % water have approximately the same boiling heat transfer coefficient as pure R-113. This result is similar to the findings of Mori et al. for oil in water emulsions using Tween[TM] 80[3] emulsifier. Ostrovskiy does not note any difference in the degree of superheat required to initiate boiling. He also investigates emulsions of water and butyl alcohol. Butyl alcohol is partially soluble in water, and mixing the two liquids produces two water-butyl alcohol solutions with different densities but the same saturation temperature. He finds that there is little difference in the boiling heat transfer coefficient for the less dense solution, the more dense solution, and emulsions of the two solutions.

Finally, Ostrovskiy investigates emulsions of water and benzene. The boiling heat transfer coefficients for these emulsions are shown in Fig. 2.4. An interesting feature of these boiling curves is that their slope for the emulsions is much lower than for pure liquids. These results are unlike those of other emulsions studied by Ostrovskiy, in which both the emulsions and the pure liquids showed similar dependence of heat transfer coefficient on heat flux. For these benzene and water emulsions, $h \propto (q'')^{0.3}$ which is close to the behavior of single-phase turbulent free convection $\left(h \propto (q'')^{0.25} \right)$. Ostrovskiy attributes this behavior to the fact that, at the heated wall, heat is removed by convection to the water, and the water is stirred by boiling of superheated benzene droplets in a turbulent-like manner.

---

[3] Tween 80 is a trademark of ICI Americas. Also known as Polysorbate 80, it is a nonionic surfactant. Span 80 (Sorbitan monooleate) is also a nonionic surfactant, and is a trademark of Croda International PLC.

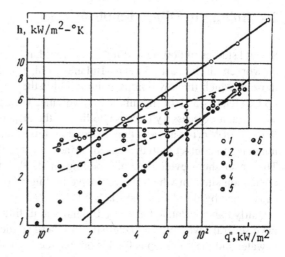

Fig. 2.4 Heat transfer coefficient for boiling water-benzene emulsions. (*1*) water; (*2*) benzene; (*3–7*) 20, 40, 55, 70, and 80 % benzene, respectively (Ostrovskiy 1988)

It is noteworthy that the portions of the boiling curves in Fig. 2.4 that follow the $h \propto (q'')^{0.3}$ curve all correspond to temperature differences of less than 20 °C. Ostrovskiy does not report the bulk temperature of the emulsion. It is reasonable to assume that the bulk of the emulsion could not have become significantly super-heated, else Ostrovskiy would have observed the "sudden foaming in the bulk" reported by Mori et al. Under this assumption, the temperature of the heated surface must be less than 20 °C above the saturation temperature of the benzene. Recalling that the data of Mori et al. indicate that superheats of greater than 20 °C are often needed to initiate boiling, it is possible that the boiling curves for water–benzene emulsions reported by Ostrovskiy resemble those for single-phase free convection because there was not, in fact, any boiling taking place. This would also explain the increase in the slope of the boiling curves for the 55, 70, and 80 % benzene emulsions at high heat flux. For all three curves, the slope of the curve changes at a temperature difference of approximately 20 °C, which is where the water in the emulsion would be expected to begin boiling as well.

That such a simple point as whether or not boiling was taking place is unclear highlights some of the difficulties in investigating boiling emulsions. The large degree of superheat required to initiate boiling means that a surface having a temperature significantly higher than the saturation temperature of the low boiling point liquid is not sufficient to ensure that boiling takes place. The opacity of emulsions generally prevents direct observation of the heated surface, unless special effort is made in the design of the test apparatus. Mori et al. watched for bubbles on the free surface of the emulsion above the heated wire to indicate boiling. The design of Ostrovskiy's apparatus may not have permitted similar observations.

2.4 Boiling in Dilute Emulsions

Several studies of boiling in dilute emulsions have been performed by Bulanov and co-workers in the past 25 years (Bulanov et al. 1984, 2006). Dilute emulsions are those in which the low boiling point liquid is the dispersed component and makes up less than ~ 5 % of the mixture by volume. They find that very high degrees of superheat are necessary to cause boiling of the dispersed component, similar to the findings of Mori et al., and they reason that this is due to the fact that the droplets, not being in contact with the heated surface, must undergo spontaneous nucleation. They also show evidence for a boiling mode that they call chain activation of nucleation sites, in which the explosive boiling of one highly superheated droplet causes nearby droplets to boil as well.

Early experimental studies by Bulanov et al. find that boiling dilute emulsions have several favorable characteristics. In these studies the low boiling point liquid is water and the continuous fluid is oil that has high saturation temperature. In both pool and flow boiling it is found that the heat transfer coefficient is always higher for the emulsion than for the pure oil. In flow boiling experiments the improvement in heat transfer is found to increase with increasing mass fractions of water, up to 33 % (Bulanov et al. 1993). In these cases the single-phase heat transfer coefficient for the emulsions is slightly higher than that of pure oil, which is explained by the higher thermal conductivity of water as compared to the oil. The heat transfer coefficient of the boiling emulsions increases uniformly with temperature until at a superheat of 80 °C the emulsion has heat transfer coefficient 1.5–2 times that of the oil alone. At moderate degrees of superheat the heat transfer coefficient of the emulsion is still lower than that of pure water undergoing boiling, but the pure water undergoes a transition to film boiling at a superheat of ~ 40 °C and has very poor heat transfer thereafter while the emulsions do not exhibit any transition to film boiling.

In pool boiling experiments with water in oil emulsions, the emulsion also shows improvement in heat transfer coefficient over a wide range of surface temperatures, up to 135 °C above the saturation temperature of water (Fig. 2.5) (Bulanov and Gasanov 2008). The improvement in heat transfer is even greater, up to ~ 3.5 times that of the oil at the highest temperature. However, the heat transfer coefficient depends on the mass fraction of water only for fractions up to 1 %. Above that limit, the heat transfer coefficient is independent of mass fraction to at least 8 % (Bulanov et al. 1996). Bulanov et al. speculate that emulsions do not undergo transition to film boiling because the boiling of dispersed droplets generates insufficient vapor to maintain the vapor film layer (Bulanov et al. 1996). In the same experiment Bulanov et al. investigate emulsions with surfactants and find ambiguous effects, which is similar to the findings of Mori et al. (1978).

One unfavorable characteristic identified in dilute emulsion boiling is the high degree of superheat required to initiate boiling of the dispersed component, often 60 °C or more (Fig. 2.5). In one study, however, the water is found to boil with very little superheat (Bulanov et al. 1984). In this case only, at low superheat the

Fig. 2.5 Pool boiling heat
transfer coefficients for (*1*)
water, (*2*) R-113, (*3*)
transformer oil, (*4*) water in
oil emulsion, and (*5*) R-113
in water emulsion (Bulanov
and Gasanov 2008)

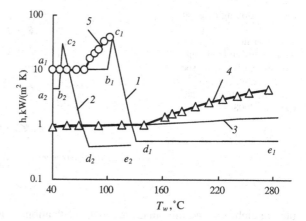

heat transfer coefficient is sometimes reduced below that of the oil. This suggests
that the suddenness of the boiling of highly superheated droplets is essential to the
heat transfer improvement, possibly because this action causes turbulence in the
surrounding liquid. The authors do not find a reason for the abnormally early onset
of boiling in this study.

Bulanov et al. explore several other aspects of boiling dilute emulsions,
including the effects of the droplet size and the addition of surfactants and suspended particles. They find that the degree of superheat required for boiling
decreases with increasing droplet size in the emulsion, although once boiling is
initiated the heat transfer coefficient has little dependence on the droplet size
(Fig. 2.6) (Bulanov and Gasanov 2007). The addition of particles may advance or
retard the onset of boiling, depending on the interaction of the particles with the
low boiling point liquid and dissolved gases. Generally, adding carbon particles to
emulsions in which water is the low boiling point liquid reduces boiling delay
owing to the tendency of the carbon to attract non-condensable gases. On the other
hand, carbon particles may increase the boiling delay when the low boiling point
liquid is an organic substance, such as pentane or ether, because the carbon also
absorbs the liquid (Bulanov and Gasanov 2008). Surfactants delay the onset of
boiling further, as shown in Fig. 2.7. They speculate that the increased delay
occurs because the surfactant coats the surface of any suspended particle as well as

Fig. 2.6 Pool boiling heat
transfer coefficients for water
in PES-5 oil emulsions with
1.0 % water by volume,
$T_\infty = 60\ ^\circ\text{C}$, and (*1*)
$d_d = 1.5\ \mu\text{m}$ and (2)
$d_d = 35\ \mu\text{m}$ (Bulanov and
Gasanov 2006)

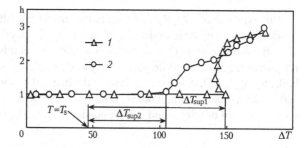

Fig. 2.7 Boiling delay at
$T_\infty = 36\ ^\circ\text{C}$ for (*1*) pure oil;
and water in oil emulsions
with (*2*) trisodiumphosphate
emulsifier, (*3*) caustic soda
emulsifier, and (*4*) no
emulsifiers (Bulanov and
Gasanov 2008)

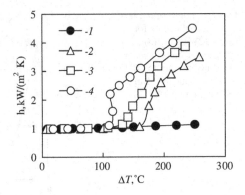

bubble embryos and thus interferes with nucleation. However, these explanations do not account for all of the experimental data (Bulanov et al. 2006).

Based on experimental results, Bulanov (2001) develops a model of boiling dilute emulsions. He assumes that each droplet that boils does so randomly, as in spontaneous nucleation, while inside the thermal boundary layer surrounding a heated surface. Boiling does not depend on contact with the heated surface itself. When a droplet boils it rises out of the thermal boundary layer owing to buoyancy, and its place is taken by more emulsion (Fig. 2.8). The probability of boiling is based on the nucleation rate of the low boiling point liquid in the emulsion, J_{eff}, and the time of residence of the droplet in the thermal boundary layer. The residence time is a function of the rate at which droplets boil and the thickness of the thermal boundary layer. For the purpose of calculating the nucleation rate, Bulanov assumes that the fluid temperature in the thermal boundary layer is uniform and equal to the heated surface temperature, and to calculate the thickness of the boundary layer he assumes a linear temperature profile. It is also assumed that the energy required to boil the droplets is provided by the liquid surrounding them, so that the temperature in the boundary layer decreases as droplets boil. These assumptions are inconsistent with each other and make a detailed examination of the behavior of the emulsion in the boundary layer problematic.

In a monodisperse emulsion in which droplets enter the boundary layer as saturated liquid and bubbles exit as saturated vapor, the heat transfer to the boiling droplets is,

$$Q = A\delta_t \varepsilon_d \rho_d i_{\text{fg}}[1 - \exp(-J_{\text{eff}}V_d\tau)]. \qquad (2.15)$$

The expression $-J_{\text{eff}}V_d\tau$ represents the probability that a droplet boils during residence time τ in the boundary layer, where J_{eff} is the effective nucleation rate and V_d is the volume of a droplet. For a monodisperse emulsion, this probability also represents the fraction of droplets, and therefore the fraction of droplet liquid, that boils. In Eq. (2.15), δ_t is the thickness of the thermal boundary layer, A is the heated surface area, and ε_d is the volume fraction of droplets, so that $A\delta_t\varepsilon_d$ represents the total volume of droplet liquid in the boundary layer. The boiling heat transfer rate is

Fig. 2.8 Configuration of Bulanov's model of boiling in an emulsion (Roesle and Kulacki 2010a)

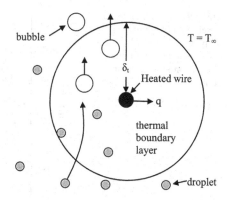

obtained by dividing Eq. (2.15) by the droplet residence time τ. The boiling heat transfer coefficient may therefore be expressed as $h = Q/A\tau(T_s - T_{sat})$.

Next, Bulanov equates the droplet residence time to the time required for a bubble to rise through the thermal boundary layer due to buoyancy. Assuming that the bubbles are small enough to be subjected to Stokes drag and that the surrounding liquid is motionless, the bubble rise time is,

$$\tau = \frac{\delta_t}{Ar} \frac{18 \, d_b \rho_d}{\mu_{eff}}, \tag{2.16}$$

where Ar is the Archimedes number of the bubble. Using this characteristic time in the equation for the heat transfer coefficient, Bulanov expresses the boiling Nusselt number as,

$$Nu = \frac{hd_b}{k_{eff}} = A_0 \frac{ArPr}{St} \varepsilon_d [1 - \exp(-J_{eff} V_d \tau)], \tag{2.17}$$

where the bubble diameter is the characteristic length. The factor of 18 in Eq. (2.16) is absorbed into an adjustment parameter A_0, which must be obtained from experiments. The Prandtl number in Eq. (2.17) is termed the mixed Prandtl number, as it contains both droplet and emulsion properties, $Pr = \mu_{eff} c_{p,d}/k_{eff}$.

Equations (2.16) and (2.17) cannot be solved yet because the boundary layer thickness in Eq. (2.16) is unknown. The boundary layer thickness is estimated by assuming that heat transfer through the boundary layer occurs by conduction and that the temperature profile is linear. Under these assumptions, the conductive heat flux through the boundary layer is $k_{eff}(T_s - T_{sat})/\delta_t$ and is set equal to the convective heat flux expressed in terms of the Nusselt number, $k_{eff} Nu(T_s - T_{sat})/d_b$, with the result $\delta_t = d_b/Nu$. The characteristic time can therefore be expressed,

$$\tau = \frac{18 \, d_b^2 \rho_d}{ArNu\mu_{eff}}. \tag{2.18}$$

Equations (2.17) and (2.18) can be solved iteratively for the heat transfer coefficient and nucleation rate J_{eff}.

Fig. 2.9 Nucleation rate J of water (*1*) and emulsified droplets of R-113 (*2*), water (*3*), pentane (*4*), and ethanol (*5*). (Bulanov and Gasanov 2008)

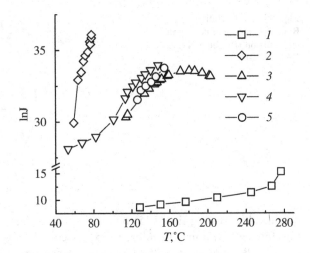

Equation (2.17) indicates that the boiling heat transfer coefficient varies with the droplet volume fraction, but experimental data show that for pool boiling the heat transfer coefficient is largely independent of droplet fraction for fractions above ∼1 %. Bulanov explains this discrepancy by noting that when a super-heated droplet boils, the energy required to vaporize the droplet comes from the surrounding liquid, so boiling droplets cause the average temperature of the emulsion to decrease. If the droplet fraction is large enough, the temperature of the emulsion will fall to the saturation temperature of the droplet liquid before all the droplets boil. For a given volume of emulsion, V, in the thermal boundary layer, the enthalpy required to vaporize all of the droplets in the volume is $V i_{fg} \rho_d \varepsilon_d$ and the sensible enthalpy that may be used to vaporize the droplets is $V\left[c_{p,c}\rho_c(1 - \varepsilon_d) + c_{p,d}\rho_d \varepsilon_d\right](T_s - T_{sat})$. The maximum droplet fraction for which all the droplets are completely vaporized, $\varepsilon_{d,0}$, can be found by equating these two enthalpies to give,

$$\varepsilon_{d,0} = \frac{c_{p,c}\rho_c(T - T_{sat})}{\rho_d i_{fg} + (c_{p,c}\rho_c - c_{p,d}\rho_d)(T - T_{sat})} \tag{2.19}$$

Bulanov uses $\varepsilon_{d,0}$ in Eq. (2.17) in place of the actual droplet fraction except for very dilute emulsions.

This model is employed not to predict heat transfer rates, but to determine the values of the parameters A_0 and J_{eff} based on experimental data. Values found for A_0 vary between 0.046 and 0.038 (Bulanov 2001; Bulanov and Gasanov 2007, 2008), and J_{eff} changes with temperature and other experimental parameters (Fig. 2.9). The factor A_0 is generally found to have a single value for a given set of experimental data. One striking result is that the nucleation rate for an emulsified liquid is much higher than for the liquid alone (compare curves 1 and 3 in Fig. 2.9).

Bulanov and Gasanov theorize that the elevated nucleation rate is due to chain activation of nucleation sites and claim that nucleation in the droplets occurs on floccules of particles with size of $O(100$ nm$)$. The floccules contain gases that were adsorbed from the surrounding liquid when they were at low temperature. When the droplet enters the thermal boundary layer, the gas is desorbed from the floccules and can serve as a nucleation site. When one highly superheated droplet boils, it does so explosively and generates a shockwave that travels for some distance through the surrounding liquid. If a second droplet is close enough, as the shockwave passes it causes floccules within the droplet to break up, thereby liberating some of their absorbed gas and causing nucleation in that droplet as well. Thus a chain boiling mechanism is set up. More recently, the conditions under which chain boiling can occur by this method have been determined by analogy of a boiling droplet to the detonation of the same mass of TNT and using the theory of point explosions (Bulanov et al. 2009). Like the overall boiling model, this analogy also assumes the formation of a shock wave by the boiling of droplets that causes breakup of floccules of nanoparticles. This condition is then used to determine the temperature at which rapid boiling by chain activation of nucleation sites first occurs (Bulanov and Gasanov 2011). The predicted trend of less superheat being required to initiate boiling for emulsions with higher dispersed component volume fraction does agree qualitatively with experiment.

Data accumulated by Bulanov and co-workers indicates that some boiling mechanism must exist aside from spontaneous nucleation, but some of the central components of their ad hoc theory of chain boiling (the breakup of floccules within droplets due to a shockwave formed by a boiling droplet) have so far not been directly shown to occur (Bulanov and Gasanov 2005, 2007). Some of the assumptions in the model are contradictory, e.g., the different temperature distributions in the boundary layer. Further, some choices of scaling in the model may not be applicable in many conditions. For example, scaling the droplet residence time using the rise velocity of individual bubbles (Eq. 2.16) neglects the overall motion of emulsion through the boundary layer due to buoyancy of the heated continuous component, which may be dominant for small droplet sizes. Therefore, even though Eqs. (2.17) and (2.18) can be fitted to experimental data, doing so does not provide insight into physics of the boiling process.

References

Behzadi A, Issa RI, Rusche H (2004) Modelling of dispersed bubble and droplet flow at high phase fractions. Chem Eng Sci 59:759–770

Blander M, Katz JL (1975) Bubble nucleation in liquids. AIChE J 21:833–848

Bouré JA, Delhaye JM (1982) General equations and two-phase flow modeling. In: Hetsroni G (ed) Handbook of multiphase systems. Hemisphere, Washington D.C

Buivid MG, Sussman MV (1978) Superheated liquids containing suspended particles. Nature 275:203–205

Bulanov NV (2001) An analysis of the heat flux density under conditions of boiling internal phase of emulsion. High Temp 39:462–469

Bulanov NV, Gasanov BM (2005) Experimental setup for studying the chain activation of low-temperature boiling sites in superheated liquid droplets. Colloid J 67:531–536

Bulanov NV, Gasanov BM (2006) Characteristic features of the boiling of emulsions having a low-boiling dispersed phase. J Eng Phys Thermophys 79:1130–1133

Bulanov NV, Gasanov BM (2007) Special features of boiling of emulsions with a low-boiling dispersed phase. Heat Transf Res 38:259–273

Bulanov NV, Gasanov BM (2008) Peculiarities of boiling of emulsions with a low-boiling disperse phase. Int J Heat Mass Transf 51:1628–1632

Bulanov NV, Gasanov BM (2011) Dependence of the beginning of chain activation of boiling sites on superheating of droplets in the dispersed phase of the emulsion. High Temp 49:213–216

Bulanov NV, Skripov VP, Khmyl'nin VA (1984) Heat transfer to emulsion with superheating of its disperse phase. J Eng Phys 46:1–3

Bulanov NV, Skripov VP, Khmylnik VA (1993) Heat transfer to emulsion with a low-boiling disperse phase. Heat Transf Res 25:786–789

Bulanov NV, Skripov VP, Gasanov BM, Baidakov VG (1996) Boiling of emulsions with a low-boiling dispersed phase. Heat Transf Res 27:312–315

Bulanov NV, Gasanov BM, Turchaninova EA (2006) Results of experimental investigation of heat transfer with emulsions with low-boiling disperse phase. High Temp 44:267–282

Bulanov NV, Gasanov BM, Muratov GN (2009) Critical volume and chain activation of boiling sites in emulsions with low-boiling dispersed phase. High Temp 47:864–869

Chen T, Klausner JF, Garimella SV, Chung JN (2006) Subcooled boiling incipience on a highly smooth microheater. Int J Heat Mass Transf 49:4399–4406

Drew DA, Passman SL (1999) Theory of multicomponent fluids. Springer, New York

Gidaspow D (1994) Multiphase flow and fluidization. Academic press, Boston

Gosman AD, Lekakou C, Politis S et al (1992) Multidimensional modeling of turbulent two-phase flow in stirred vessels. AIChE J 38:1946–1956

Hao YL, Tao YX (2003a) Non-thermal equilibrium melting of granular packed bed in horizontal forced convection part I: experiment. Int J Heat Mass Transf 46:5017–5030

Hao YL, Tao YX (2003b) Non-thermal equilibrium melting of granular packed bed in horizontal forced convection part II: numerical simulation. Int J Heat Mass Transf 46:5031–5044

Hao YL, Tao YX (2004) A numerical model for phase-change suspension flow in microchannels. Numer Heat Transf Part A: Appl 46:55–77

Hill DP (1998) The computer simulation of dispersed two-phase flows. Ph.D. Thesis, University of London, London

Iida Y, Okuyama K, Sakurai K (1994) Boiling nucleation on a very small film heater subjected to extremely rapid heating. Int J Heat Mass Transf 37:2771–2780

Ishii M, Zuber N (1979) Drag coefficient and relative velocity in bubbly, droplet or particulate flows. AIChE J 25:843–855

Kagan Y (1960) The kinetics of boiling of a pure liquid. Russ J Phys Chem 34:42–46

Katz JL, Blander M (1973) Condensation and boiling: corrections to homogeneous nucleation theory for nonideal gases. J Colloid Interface Sci 42:496–502

Mori YH, Inui E, Komotori K (1978) Pool boiling heat transfer to emulsions. J Heat Transf Trans ASME 100:613–617

Mori YH, Sano H, Komotori K (1980) Cinemicrophotographic study of boiling of water-in-oil emulsions. Int J Multiphase Flow 6:255–266

Ostrovskiy NYu (1988) Free-convection heat transfer in the boiling of emulsions. Heat Transf-Sov Res 20:147–153

Pavlov PA, Skripov VP (1970) Explosive boiling of liquids and fluctuation nucleus formation. High Temp 8:833–839

Pope SB (2000) Turbulent flows. Cambridge University Press, Cambridge

Roesle ML, Kulacki FA (2010) Boiling of dilute emulsions—toward a new modeling framework. Ind Eng Chem Res 49:5188–5196

Rusche H (2002) Computation fluid dynamics of dispersed two-phase flows at high phase fractions. Ph.D. Thesis, University of London, London

Rusche H, Issa RI (2000) The effect of voidage on the drag force on particles, droplets, and bubbles in dispersed two-phase flow. In: Proceedings of the Japanese European two-phase flow group meeting, Tsukuba, Japan

Skripov VP (1974) Metastable liquids. Wiley, New York

Soo SL (1967) Fluid dynamics of multiphase systems. Blaisdell, Waltham

Wakeshima H, Takata K (1958) On the limit of superheat. J Phys Soc Jpn 13:1398–1403

Wallis GB (1969) One-dimensional two-phase flow. McGraw-Hill, New York

Wismer KL (1922) The pressure-volume relation of superheated liquids. J Phys Chem 22:301–315

Xing KQ, Tao YX, Hao YL (2006) Slurry viscosity study and its influence on heat transfer enhancement effect of PCM slurry flow in micro/mini channels. In: Proceedings of the 2006 ASME international mechanical engineering congress and exposition, New York

Chapter 3
Boiling of a Single Droplet

Keywords Bubble growth · Droplet evaporation · Explosive evaporation · Surface tension · Inertia-dominated growth · Thermal diffusion · Bubble oscillation · Numerical model

To understand the overall behavior of boiling emulsions, it is helpful to first understand what happens when a single highly superheated droplet boils. Some key questions are how long the boiling process takes, how the bubble behaves after all the liquid in the droplet has evaporated, and what effect the boiling droplet has on the surrounding liquid.

The study of growing vapor bubbles in liquid started with bubbles in an infinite domain of uniformly superheated liquid. In these studies the bubble is generally assumed to be spherical and stationary in the surrounding liquid. Expansion of the bubble is initially limited by the inertia of the liquid and later by thermal diffusion to the bubble surface. More recently boiling in suspended droplets with $d_d \sim 1$ mm has been studied, usually in the context of direct contact heat exchangers. Buoyancy becomes important in these cases because the growing vapor bubble rises through the liquid with the evaporating droplet suspended from the bottom of the bubble. The rate of boiling in these cases is determined by convection to the bubble-droplet pair as they rise through the surrounding liquid.

Droplets that make up the emulsions considered here are two to three orders of magnitude smaller than those found in direct contact heat exchangers, and so their behavior during boiling is also quite different. Such small droplets have insignificant drift velocity and boil so quickly that gravitational effects are negligible. And unlike the bubble expanding in superheated liquid, the heat transfer properties of the liquid around the droplet are important in the later stages of boiling. The behavior of the bubble after the droplet has completely evaporated also merits examination. Roesle and Kulacki (2010) have presented a summary of the more complete analysis presented in this chapter.

M. L. Roesle and F. A. Kulacki, *Boiling Heat Transfer in Dilute Emulsions*,
SpringerBriefs in Thermal Engineering and Applied Science,
DOI: 10.1007/978-1-4614-4621-7_3, © The Author(s) 2013

3.1 Prior Work

The growth of vapor bubbles in a superheated liquid involves the effects of thermal diffusion and inertia in the surrounding liquid as well as surface tension at the surface of the bubble. Different effects are dominant at different times and under different conditions and accurate modeling of all stages of bubble growth can be accomplished only with numerical simulations. Boiling of large suspended droplets is more complex still, as buoyancy causes the droplet and bubble to move through the surrounding liquid and lose spherical symmetry.

Generally a vapor bubble that forms in superheated liquid undergoes three phases of growth. The bubble starts with the same temperature as the surrounding liquid and its growth is driven by the difference between the vapor pressure inside the bubble and the ambient pressure. Immediately after the bubble first forms, this driving pressure difference is nearly balanced by the surface tension of the bubble and the bubble experiences surface tension dominated growth (Fig. 3.1a). The pressure jump due to surface tension varies with the inverse of the bubble diameter so that the effect of surface tension quickly becomes negligible as the bubble size increases. After the pressure jump due to surface tension becomes small, the rate of bubble growth is limited by the inertia of the surrounding liquid (Fig. 3.1b). As the bubble continues to expand, evaporation at its surface causes its temperature to decrease. The vapor pressure in the bubble decreases until eventually there is

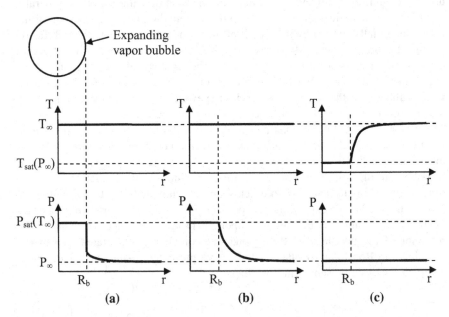

Fig. 3.1 Pressure and temperature fields in and around vapor bubbles for **a** surface tension-dominated growth, **b** inertia-dominated growth, and **c** thermal diffusion-dominated growth (Roesle and Kulacki 2010)

negligible driving force due to pressure. After this point growth is limited by the rate of heat diffusion to the bubble surface (Fig. 3.1c). For a bubble growing in uniformly superheated liquid, this phase of growth continues indefinitely (Lee and Merte 1996).

If surface tension is neglected, growth dominated by inertia and thermal diffusion in a uniformly superheated liquid can be described analytically (Mikic et al. 1970),

$$R^+ = \frac{2}{3}\left[(t^+ + 1)^{3/2} - (t^+)^{3/2} - 1\right], \tag{3.1}$$

where the dimensionless bubble radius, R^+, and time, t^+, are given by,

$$R^+ = \frac{\pi}{12}\frac{R}{\mathrm{Ja}^2\alpha_f}\left(\frac{2i_{\mathrm{fg}}\rho_v\Delta T}{3\rho_f T_{\mathrm{sat}}}\right)^{1/2}, \tag{3.2}$$

$$t^+ = \frac{\pi}{18}\frac{t}{\mathrm{Ja}^2\alpha_f}\frac{i_{\mathrm{fg}}\rho_v\Delta T}{\rho_f T_{\mathrm{sat}}} \tag{3.3}$$

This equation approaches the solution of Plesset and Zwick (1954) for thermal diffusion dominated growth at large t^+ ($R^+ = (t^+)^{1/2}$). At small t^+, the bubble radius increases linearly with time ($R^+ = t^+$), similar to the Rayleigh solution for inertia dominated growth (Rayleigh 1917). Unlike the Rayleigh solution, however, Eqs. (3.2) and (3.3) do not contain the driving pressure difference. Instead the pressure difference is replaced with the driving temperature difference $\Delta T = T_\infty - T_{\mathrm{sat}}(P_\infty)$ through the use of a linear approximation of the Clausius-Clapeyron equation. This approximation can result in under prediction of the bubble growth rate for large degrees of superheat (Theofanous and Patel 1976), but Eq. (3.1) is found to agree well with experimental data for moderate superheat (Lien 1969).

Analytical solutions are generally not possible when more accurate models of fluid properties or the thermal boundary layer are utilized. The numerical study by Lee and Merte (1996) uses accurate models of fluid properties and simulates the temperature field surrounding the bubble. They also include the surface tension of the bubble in their model. They begin their simulations with a bubble radius very close to R_{cr} (Eq. 2.1) under isothermal conditions. Because a bubble at the critical radius is in unstable equilibrium, a small perturbation in the bubble radius is required to cause the bubble to expand. Lee and Merte show that for small perturbations, the magnitude of the perturbation has no effect on the subsequent expansion of the bubble aside from a small change in the time that elapses before the bubble begins to expand rapidly. Their simulations show good agreement with experimental studies for a wide range of fluids, ambient pressures, and degrees of superheat. They also find that the earlier model of Mikic et al. is accurate in many cases except for the early surface tension dominated growth.

These studies of bubbles in uniformly superheated liquid have limited applicability to suspended droplets because they do not consider the interaction between the bubble, droplet, and the surrounding liquid, as well as the loss of spherical symmetry that occurs due to buoyancy. A relation between the Nusselt and the Péclet numbers of an evaporating droplet was developed by Sideman and Taitel (1964) based upon a simple analytical model. They assume that the bubble is spherical with the droplet forming a layer around the bottom portion of the bubble and steady potential flow around the bubble and droplet as they rise due to buoyancy. They find good agreement with experimental results for evaporating droplets of pentane and butane with $1.9 < d_d < 3.9$ mm. Tochitani and co-workers (Tochitani et al. 1977a, b) develop a similar model assuming Stokes flow around the droplet. They also determine a relationship between the Nusselt and the Péclet numbers and find agreement with experimental results for $0.8 < d_d < 1.4$ mm and Re < 1. Battya et al. (1984) reexamine the data of Sideman and Taitel and find that the Nusselt number also depends on the Jakob number. The correlations of Sideman and Taitel, Tochitani et al., and Battya et al. all predict that Nu $\rightarrow 0$ as Pe $\rightarrow 0$, and thus none of them are appropriate for very small droplets that have negligible velocity relative to the surrounding liquid.

Other researchers have examined boiling droplets using more sophisticated models of the geometry of the bubble and droplet (Vuong and Sadhal 1989a, b). Raina and Grover (1985) investigate the effects of sloshing of the droplet around the bubble. Mahood (2008) performs numerical simulations of boiling droplets in which the bubble and droplet are concentric, and Wohak and Beer (1998) perform axisymmetric simulations of a boiling deformable droplet. All of these studies examine droplets with $d_d \sim 1$ mm that boil over time scales of milliseconds to seconds. Therefore, they address a different set of processes and forces than are dominant in boiling of very small droplets.

Shepherd and Sturtevant (1982) address rapid boiling of a superheated droplet. They observe boiling in droplets with $0.5 < d_d < 1$ mm near their limit of superheat and find that the vapor bubble grows more rapidly than predicted for thermal diffusion growth when the bubble reaches approximately the original size of the droplet. They attribute this deviation from theory to instabilities that roughen the bubble surface. They also observe oscillation of the resulting vapor bubble and instabilities that occur there as well. These instabilities occur only for droplets much larger than those found in emulsions. Lee and Merte (2005) and Frost and Sturtevant (1986) also address instability of large boiling droplets near the superheat limit.

Kwak et al. (1995) consider the oscillating bubble resulting from a droplet boiling at its superheat limit. Their coupled differential equations for the evolution over time of the bubble pressure, temperature, radius and velocity, and the thickness of the thermal boundary layer are,

$$\frac{dP_b}{dt} = -\frac{3\gamma P_b}{R_b}\frac{dR_b}{dt} - \frac{6(\gamma - 1)k_c(T_b - T_\infty)}{\delta_t R_b}, \tag{3.4}$$

$$\frac{dT_b}{dt} = \frac{3(\gamma - 1)T_b}{R_b}\frac{dR_b}{dt} - \frac{6(\gamma - 1)k_c T_b(T_b - T_\infty)}{\delta_t R_b P_b}, \tag{3.5}$$

$$R_b \frac{du}{dt} + \frac{3}{2} u^2 = \frac{1}{\rho_c} (P_b - P_\infty), \tag{3.6}$$

$$u = \frac{dR_b}{dt}, \tag{3.7}$$

$$\left[1 + \frac{\delta_t}{R_b} + \frac{3}{10} \left(\frac{\delta_t}{R_b} \right)^2 \right] \frac{d\delta_t}{dt} = \frac{6\alpha_c}{\delta_t} - \left[\frac{2\delta_t}{R_b} + \frac{1}{2} \left(\frac{\delta_t}{R_b} \right)^2 \right] \frac{dR_b}{dt}$$
$$- \delta_t \left[1 + \frac{\delta_t}{2R_b} + \frac{1}{10} \left(\frac{\delta_t}{R_b} \right)^2 \right] \frac{1}{T_b - T_\infty} \frac{dT_b}{dt}. \tag{3.8}$$

They assume a quadratic temperature profile in the boundary layer. Because the last term of Eq. (3.8) contains the temperature difference $(T_b - T_\infty)$ in the denominator, they set $d\delta_t/dt = 0$ when $|T_b - T_\infty| < 0.21\,°C$. This restriction ensures that the boundary layer thickness remains small compared to the bubble radius. They do not address in detail the boiling process that leads to the oscillating bubble. Park et al. (2005) compare this model to measurements for droplets of different hydrocarbons with $d_d \sim 1$ mm boiling at their superheat limit, and also develop a model for the pressure field far from the droplet during boiling. Park et al. add viscous and surface tension terms to Eq. (3.6) although, for the size of the bubble they simulate, surface tension effects should be negligible.

3.2 The Boiling Model

We develop a model for the boiling of a single droplet that accounts for droplet size, degree of superheat, and combinations of fluid parameters for the droplet and surrounding liquid. The simulations include thermal diffusion through the liquid surrounding the vapor bubble and the momentum of the liquid. A key simplifying assumption is that the bubble nucleus forms and remains at the center of the spherical droplet so that spherical symmetry is maintained (Fig. 3.2a). Each liquid is assumed to be incompressible and have constant properties. The bubble vapor is assumed to obey the ideal gas law, and the temperature and pressure are assumed to be uniform throughout the bubble.

A more realistic model would allow for departures from spherical symmetry if the bubble nucleus forms at some point away from the center of the droplet (Fig. 3.2b). In this case, the droplet retains radial symmetry around an axis defined by the centers of the bubble and the droplet. Such an axisymmetric model is much more complex than the one-dimensional model described here, but the one-dimensional model is expected to be reasonably accurate. An early analytical model of vapor bubble growth in uniformly superheated liquid found that the departure from symmetry caused by the presence of a solid wall adjacent to the bubble has little impact on bubble growth, slowing the rate of growth by about

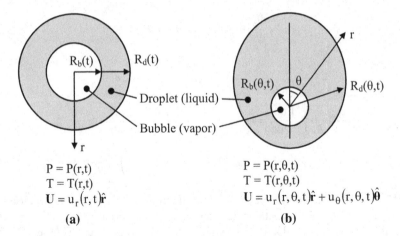

Fig. 3.2 Models of a boiling droplet. **a** Spherically symmetric one-dimensional. **b** Axisymmetric two-dimensional model

one-third but not changing the form of the resulting equations (Mikic et al. 1970). It is assumed that departures from spherical symmetry in small droplets would have similarly small impact.

The boiling process can be divided into two time domains based on the presence of droplet liquid. In the first time domain, the growth of the vapor bubble is accompanied by evaporation of the droplet liquid at the bubble surface. During this period the vapor in the bubble is assumed to be saturated, and the rate of evaporation of liquid at the droplet surface must be taken into account. The second time domain begins when the liquid has completely evaporated. During this period the mass of the vapor bubble remains constant and the vapor may become superheated. Expansion is assumed to be uniform and thus the liquid velocity surrounding the bubble has only a radial component. The state of the vapor in the bubble is determined using the ideal gas law and the Clausius-Clapeyron equation.

Recalling that the problem has spherical symmetry and that the liquids around the bubble are incompressible, the velocity in the liquid can be expressed,

$$u(r, t) = \frac{F(t)}{r^2}. \tag{3.9}$$

If the velocity of the liquid can be found at any location at a given time step, the velocity everywhere in the liquid is known. When this equation for velocity is inserted into the momentum equation a differential equation for pressure is obtained,

$$\frac{\partial P}{\partial r} = \rho \left(\frac{2}{r} u^2 - \frac{1}{r^2} \frac{dF(t)}{dt} \right). \tag{3.10}$$

Integrating Eq. (3.10) in radius and using the limits $P = P_b$ at $r = R_b$ and $P = P_\infty$ at $r = \infty$, the equation for the evolution of $F(t)$ over time is,

$$0 = (P_b - P_\infty) - \frac{2\sigma}{R_b} + \frac{F(t)^2}{2}\left(\frac{\rho_d}{R_b^4} + \frac{\rho_c - \rho_d}{R_d^4}\right) + \frac{dF(t)}{dt}\left(-\frac{\rho_d}{R_b} + \frac{\rho_d - \rho_c}{R_d}\right).$$

$$(3.11)$$

Equation (3.11) also includes a term for the pressure jump at the bubble surface due to surface tension. It is assumed that the droplet is large enough that the pressure jump at the droplet surface is negligible. The pressure jump at each surface due to viscosity is also assumed to be negligible. Equation (3.11) also accounts for differing densities in the droplet and the surrounding liquid. The pressure in the bubble is obtained at any time using the equation of state of the vapor in the bubble.

The thermal energy balance in the liquid is,

$$\frac{\partial T}{\partial t} + u\frac{\partial T}{\partial r} = \alpha\left(\frac{\partial^2 T}{\partial r^2} + \frac{2}{r}\frac{\partial T}{\partial r}\right). \tag{3.12}$$

The temperature is assumed to be uniform inside the vapor bubble, and during the first time domain it is assumed that the vapor is saturated because it is in contact with its own liquid at the bubble surface. The saturation pressure and temperature are linked by the Clausius-Clapeyron equation, assuming that the latent heat of vaporization is constant,

$$\frac{P_{\text{sat}}}{P_{\text{sat},0}} = \exp\left[\frac{i_{\text{fg}}}{R_G}\left(\frac{1}{T_{\text{sat},0}} - \frac{1}{T_{\text{sat}}}\right)\right]. \tag{3.13}$$

After the droplet has completely evaporated, the temperature in the bubble is determined using the energy balance of the bubble. The energy balance includes the work done by the expanding bubble and heat diffusion to the bubble from the surrounding liquid,

$$\frac{dT_b}{dt} = \frac{4\pi R_b^2}{m_b c_{v,b}}\left(k\frac{\partial T}{\partial r}\bigg|_{r=R_b} - P_b\frac{dR_b}{dt}\right). \tag{3.14}$$

During the first time domain, the mass of vapor in the bubble increases due to evaporation according to,

$$\frac{dm_b}{dt} = \frac{4\pi R_b^2 k_d}{i_{\text{fg}}}\frac{\partial T}{\partial r}\bigg|_{r=R_b}. \tag{3.15}$$

Evaporation at the bubble surface also results in a relative velocity between the bubble surface and the liquid adjacent to the surface given by,

$$\frac{dR_b}{dt} - u\big|_{r=R_b} = \frac{1}{4\pi R_b^2 \rho_d}\frac{dm_b}{dt}. \tag{3.16}$$

A summary of the variables, initial and boundary conditions, and governing equations is given in Table 3.1. Two sets of governing equations are given. The first

Table 3.1 Quantities and formulas in the one-dimensional model of a boiling droplet (Roesle and Kulacki 2010)

Quantity	Initial/boundary conditions	Governing equation	
		With phase change	Without phase change
$T(r, t)$	$T(r, 0) = T_\infty$ $T(R_b, t) = T_b$	Eq. (3.12)	Eq. (3.12)
$F(t)$	$F(0) = 0$	Eq. (3.11)	Eq. (3.11)
$m_b(t)$	$m_b(0) = \frac{4}{3}\pi R_b^3(0)\frac{P_\infty}{R_G T_\infty}$	Eq. (3.15)	$m_b = m_d$
$R_b(t)$	$R_b(0) = \frac{1.001 \cdot 2\sigma}{P_{sat}(T_\infty) - P_\infty}$	Eq. (3.16)	$\frac{dR_b}{dt} = u\vert_{r=R_b}$
$R_d(t)$	N/A	$R_d = \left(R_b^3 + \frac{m_d - m_b}{4/3\pi\rho_d}\right)^{1/3}$	$R_d = R_b$
$T_b(t)$	$T_b(0) = T_\infty$	$T_b = T_{sat}$	Eq. (3.14)
$P_b(t)$	N/A	$P_b = P_{sat} = \rho R_G T_b$	$P_b = \rho R_G T_b$

set is for the first time domain in which evaporation occurs. The second set applies when the droplet is completely evaporated and the bubble is superheated, in which case no phase change occurs. The equations for the first time domain also apply to any time after the complete evaporation of the droplet in which the vapor in the bubble becomes saturated again and condensation occurs at the bubble surface.

3.3 Solution

For all solutions discussed here, the bubble begins at rest with an initial radius 0.1 % larger than R_{cr} and in thermal equilibrium with the ambient. The ambient temperature and pressure are held constant throughout each solution, and solutions are obtained with an ambient pressure of 1 atm. The solution represents the temperature field in the liquid around the bubble by a one-dimensional array of nodes. The first node is located at the surface of the bubble and represents the temperature of the bubble. It is assumed that there is no temperature jump at the bubble surface. The last node has temperature fixed at the ambient temperature. Simulations are performed with a total of sixty nodes that are spaced to span a distance of 40 μm from the bubble surface. Using an approach similar to that of Lee and Merte (1996), nodes are clustered near the bubble surface according to,

$$r_i = R_b + (40\,\mu m)\left(\frac{i}{N-1}\right)^{2.2}, \tag{3.17}$$

where i is the node number, $0 \le i \le N - 1$. As the solution progresses, the nodes move outwards with uniform velocity equal to that of the bubble surface. The first node is therefore always located at the bubble surface and the spacing between nodes remains constant.

The equations are solved by a fully implicit iterative method. The discretized form of Eq. (3.12) is,

$$T_i = \frac{a_{i,1}T_{i-1} + a_{i,2}T_{i+1} + a_i^t T_i^0}{a_{i,1} + a_{i,2} + a_i^t}. \tag{3.18}$$

The coefficients $a_{i,1}$ and $a_{i,2}$ account for energy transfer through the inner and outer surfaces, respectively, of node i, and they take into account the motion of the nodes themselves as well as that of the liquid,

$$a_{i,1} = \frac{k(r_i + r_{i-1})^2}{r_i - r_{i-1}} + \rho c \left[2F - \frac{R_b - R_b^0}{2(\Delta t)}(r_i + r_{i-1})^2 \right], \tag{3.19}$$

$$a_{i,2} = \frac{k(r_{i+1} + r_i)^2}{r_{i+1} - r_i} + \rho c \left[\frac{R_b - R_b^0}{2(\Delta t)}(r_{i+1} + r_i)^2 - 2F \right]. \tag{3.20}$$

The coefficient a_i^t accounts for temperature change within node i,

$$a_i^t = \frac{\rho c}{6(\Delta t)} \left[(r_{i+1} + r_i)^3 - (r_i + r_{i-1})^3 \right]. \tag{3.21}$$

Defining k, ρ, and c in Eqs. (3.19)–(3.21) requires some care at the boundary between the droplet and surrounding liquid. In Eqs. (3.19) and (3.20), ρ and c are set equal to the properties of the liquid present at the inner and outer surface of the node. The thermal conductivity is a weighted average of k_c and k_d that accounts for conduction through the droplet liquid and surrounding liquid in series. In Eq. (3.21), ρc is a weighted average of the droplet and surrounding liquid values based on the fraction of the node occupied by each liquid. The averages are linear averages based on radii. This approach ignores the curvature of the nodes, but the effect of this approximation is small as long as the distance between nodes is small compared to the radius of each node.

The number of iterations at each time step is typically set at fifteen, and the equations are under relaxed using a relaxation parameter between 0.6 and 0.7. The initial time step size is set to 1 ns, and is increased somewhat as the simulation progresses. Solution durations are between 50 μs and 1 ms and are always too short for the thermal boundary layer to grow to the outer edge of the simulated region of liquid. The droplets take between 10 and 100 μs to boil. Droplet radii are chosen in the range $2 \le R_d \le 15$ μm.

Droplets of water in mineral oil are simulated to investigate configurations in which the droplet liquid has much higher thermal conductivity than the surrounding liquid. For this combination of fluids superheats of 40 and 80 °C are used, corresponding to temperatures over which boiling of water in oil emulsions has been observed experimentally (Mori et al. 1978). For droplets of pentane and FC-72 suspended in water, the surrounding liquid has higher thermal conductivity and specific heat than the droplet. For these configurations smaller degrees of superheat, 20–60 °C, are considered. For all three pairs of fluids, properties are evaluated at the saturation temperature of the more volatile liquid and are assumed to be constant.

3.4 Results

The initial behavior of the vapor bubble is qualitatively similar to that observed by Lee and Merte (1996). Bubble growth accelerates as it transitions from surface tension dominated growth to inertia dominated growth, where the bubble radius increases at a constant rate (Fig. 3.3). A novel result here is that the rate of expansion of the bubble depends on the droplet size. Equation (3.11) shows that a difference in density between the droplet liquid and the surrounding liquid can impact the rate of growth of the bubble, but only when the bubble and droplet radii are comparable. When $R_b \ll R_d$ inertial effects due to any difference in density are negligible. Therefore, for water in mineral oil ($\rho_d/\rho_c = 1.17$) the rate of expansion is nearly independent of the droplet size, while for pentane in water ($\rho_d/\rho_c = 0.614$) the rate of expansion is significantly greater in larger droplets. For FC-72 in water ($\rho_d/\rho_c = 1.63$) the opposite trend is observed.

After the initial inertia dominated growth, bubble growth approaches the analytical solution of Mikic et al. (1970) until the droplet is mostly evaporated (Figs. 3.4, 3.5, 3.6). For water in oil and FC-72 in water (fluid combinations for which $\rho_d > \rho_c$) the simulated bubble growth is generally faster than predicted by Eq. (3.1), while the opposite is true for pentane in water ($\rho_d < \rho_c$). This result is explained by the deviation in the rate of bubble growth during the momentum dominated growth described in the previous paragraph. The simulated bubble growth also tends to run faster than Eq. (3.1) for larger ΔT. This discrepancy is most likely due to the relation used in this study for the saturation pressure [Eq. (3.13)] which results in a larger initial driving pressure than the relation used by Mikic et al. (1970).

As the bubble expands, the layer of droplet liquid surrounding the bubble becomes thinner and eventually evaporates completely. As it does so, the thermal boundary layer around the bubble grows into the liquid that surrounds the droplet. For the water in oil case, the oil has lower thermal conductivity, specific heat, and density than the water. All three factors contribute to reducing the rate at which

Fig. 3.3 Initial bubble growth. Water droplets in oil have $\Delta T = 40$ °C, droplets suspended in water have $\Delta T = 20$ °C. *Heavy lines* indicate droplets with $R_d = 2$ μm, *light lines* $R_d = 15$ μm

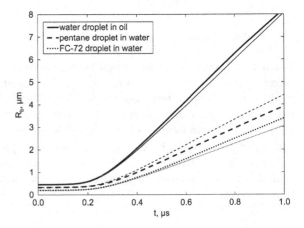

Fig. 3.4 Bubble growth for water droplets suspended in mineral oil. Equation (3.1) calculated using properties of water

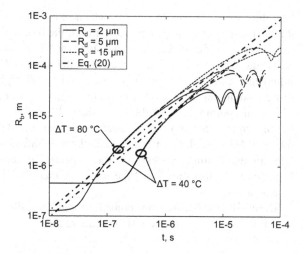

Fig. 3.5 Bubble growth for pentane droplets suspended in water. Equation (3.1) calculated using properties of pentane

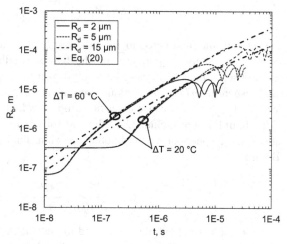

Fig. 3.6 Bubble growth for FC-72 droplets suspended in water. Equation (3.1) calculated using properties of FC-72

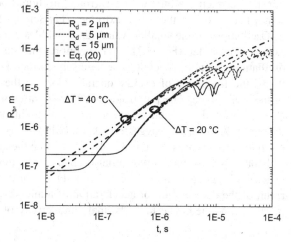

heat diffuses to the bubble surface, which reduces the rate of bubble expansion in the later stages of boiling. For water in oil, the thermal properties of the oil begin to influence the rate of bubble expansion when it reaches approximately one-third of its equilibrium radius. For pentane in water and FC-72 in water, the surrounding liquid has much higher thermal conductivity and higher specific heat than the droplet liquid, so the bubble accelerates near the end of the boiling process.

For all combinations of fluids considered here, these effects at the end of the boiling process are most pronounced for larger droplets and for lower ΔT. The model of Mikic et al. (1970) shows that thermal diffusion dominated growth occurs later in the boiling process, so the dependence on droplet size is expected. Equation (3.1) also shows that the region of inertia dominated growth grows with increasing ΔT.

Eventually the bubble must come to rest at its equilibrium radius. In the absence of significant surface tension effects this radius is,

$$R_{b,0} = R_{d,0} \left(\frac{\rho_d R_G T_\infty}{\rho_\infty} \right)^{1/3}. \tag{3.22}$$

However, the liquid around the bubble still has considerable velocity when the droplet completely evaporates, and the bubble oscillates around its equilibrium radius. The oscillations decay as a result of thermal and acoustic damping (Plesset and Prosperetti 1977). These simulations are not expected to accurately predict the decay of oscillation because acoustic damping, which depends on compressibility of the liquid, is not included. Neglecting surface tension, small amplitude oscillations of a spherical vapor bubble are sinusoidal at the Minnaert frequency (Brennan 1995),

$$f_M = \frac{1}{2\pi R_{b,0}} \left(\frac{3\kappa P_\infty}{\rho_c} \right)^{1/2}. \tag{3.23}$$

For an isothermal bubble $\kappa = 1$, and for an adiabatic bubble, $\kappa = \gamma$. When heat transfer between the bubble and surrounding liquid occurs, κ is expected to fall somewhere between these two extremes (Plesset and Prosperetti 1977).

Oscillations of the smaller water vapor bubbles are neither small amplitude nor exactly sinusoidal (Fig. 3.7), and therefore the analytical solutions for small oscillations are not expected to be precisely correct. For all three combinations of fluids, the frequency of oscillation falls close to the Minnaert frequency for an isothermal bubble. The polytropic coefficient varies from a minimum of 0.95 for $R_d = 5$ μm in the pentane in water case to a maximum of 1.06 for $R_d = 2$ μm in the water in oil case, suggesting that the bubble is close to isothermal during oscillation in all cases. Figure 3.7 also shows that the dependence of the magnitude of bubble oscillations on the initial droplet size is much more significant for the water in oil case than for the pentane and FC-72 in water cases. This behavior is a result of the deceleration or acceleration of bubble expansion when the droplet is mostly evaporated. The high thermal conductivity of water also results in much

Fig. 3.7 Oscillation of vapor bubbles resulting from droplets with **a** $R_d = 2$ μm and **b** $R_d = 15$ μm

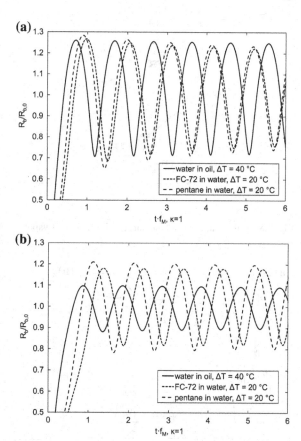

faster decrease in the magnitude of the oscillation due to thermal damping than in the water in oil case.

Figures 3.8 and 3.9 show the evolution of bubble temperature for the water in oil and FC-72 in water cases. Figure 3.8 illustrates the decrease in bubble temperature during the early expansion of the bubble. In all cases the initial decrease in bubble temperature is nearly independent of the droplet radius, which is consistent with the observation that the initial bubble expansion is also nearly independent of the droplet radius. For the smaller water droplets in oil, there is a second stage of rapid cooling that occurs after the initial expansion begins. This behavior is a result of the thermal boundary layer growing into the oil around the water droplet. As this occurs the rate of heat transfer to the bubble surface decreases and therefore the evaporation rate decreases. Inertia of the surrounding liquid prevents the rate of expansion of the bubble from changing instantly. The specific volume of the vapor in the bubble thus rises, and the temperature decreases. The FC-72 droplet in water exhibits the opposite behavior, where the temperature in the bubble begins to rise shortly before the droplet completely

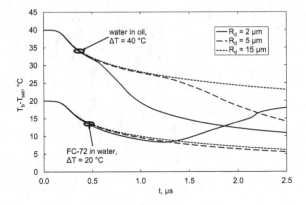

Fig. 3.8 Temperature at bubble surface during initial vapor bubble expansion

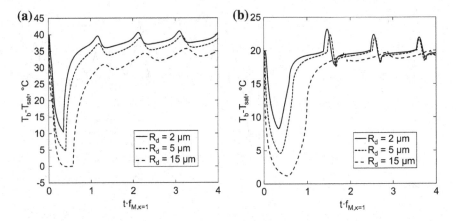

Fig. 3.9 Temperature at bubble surface for **a** water in oil, $\Delta T = 40$ °C and **b** FC-72 in water, $\Delta T = 20$ °C (Roesle and Kulacki 2010)

evaporates owing to the increase in the heat transfer rate as the layer of liquid FC-72 around the bubble becomes thin.

Figure 3.9 shows the variation in bubble temperature during the entire boiling process and the first few oscillations of the bubble. Figure 3.9a shows that only for the largest water droplet in oil does the bubble temperature ever reach T_{sat}, so that in most cases the bubble never experiences pure thermal diffusion limited growth. The figure also shows that the bubbles are indeed close to isothermal when oscillating. The sharp changes in the FC-72 bubble temperature (Fig. 3.9b) occur near the minimum radius in each oscillation and are caused by FC-72 vapor briefly becoming saturated. A small amount of FC-72 condenses onto the surface of the bubble, and acts as thermal insulation for the bubble due to its poor thermal conductivity. As the bubble expands in the next cycle of oscillation, the FC-72 evaporates again. To investigate the significance of this condensation, additional simulations were performed of FC-72 droplets in water in which condensation was suppressed. The bubble thus becomes slightly subcooled during a portion of each oscillation, but the bubble behavior is not significantly different from the cases in which condensation occurs.

The model developed by Kwak et al. (1995) is also applied to the bubbles simulated here. Because their model only describes oscillating bubbles, initial conditions for numerical calculation using Eqs. (3.4)–(3.8) are taken from the results of the simulations performed for this study at the time when the droplet evaporates completely. The initial thermal boundary layer thickness is set so that the initial heat transfer rate to the bubble is the same as the rate predicted by this study. The restriction on Eq. (3.8) is expanded to $d\delta_t/dt = 0$ when $|T_b - T_\infty|$ $< 1°C$ to ensure that the boundary layer thickness remains small compared to the bubble radius.

Figure 3.10 compares the results of this study to simulations performed using the Kwak et al. model for the oscillating bubble that results from the boiling of a droplet of FC-72 in water with $R_d = 15$ μm and $\Delta T = 40$ °C. As Fig. 3.10a illustrates, the two models are in very close agreement for the frequency and initial amplitude of the oscillations. However, Kwak et al. predict significantly larger variation in the bubble temperature, and the variation increases with time as the thermal boundary layer grows (Fig. 3.10b). The most likely reason for these discrepancies is the assumption by Kwak et al. of a quadratic temperature profile in the boundary layer. As Fig. 3.10c illustrates, the present model predicts that the temperature profile in the boundary layer can become complex. The shape of the temperature profile at $t = 40$ μs contains the entire history of the boiling droplet including heat transfer to the droplet during the boiling process and bubble expansion ($0 < t < 24$ μs), heat transfer from the bubble during its subsequent contraction ($24 < t < 38$ μs), and finally heat transfer to the bubble again as it begins to expand again (38 μs $< t$). Any model that assumes a temperature profile in the boundary layer cannot capture this behavior.

Additional simulations are performed for $1 \leq R_d \leq 30$ μm and $4 \leq \Delta T \leq 45$ °C for FC-72 droplets in water. The primary quantity of interest for these simulations is the ratio of the maximum radius achieved by the bubble to its steady-state radius, which will be used in the model of boiling emulsions developed in the next chapter. For this range of conditions, $R_{b,max}/R_b$ can be described with RMS error of 0.0026 by,

Fig. 3.10 Predictions of
a radius and **b** temperature
for oscillating FC-72 vapor
bubble in water, and
c boundary layer temperature
profile, $\Delta T = 40$ °C and
$R_d = 15$ μm (Roesle and
Kulacki 2010)

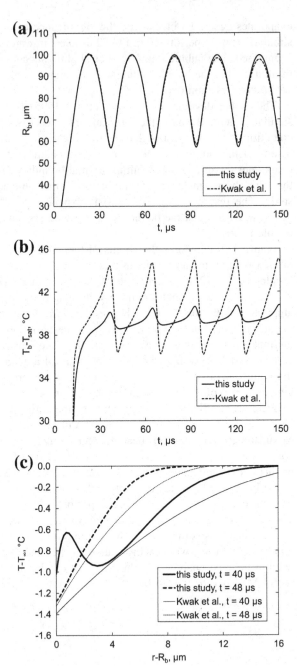

$$\frac{R_{b,\text{max}}}{R_b} = 1.74 \times 10^{-6} \left(\frac{\Delta T}{°C}\right)^3 - 1.97 \times 10^{-4} \left(\frac{\Delta T}{°C}\right)^2 + 0.011 \frac{\Delta T}{°C} + 0.9767$$

$$+ \left[0.04\ln\left(\frac{\Delta T}{°C}\right) - 0.00322\right] \exp\left(-0.0839\frac{R_d}{\mu m} + 0.3004\right).$$

$$(3.24)$$

This analysis shows also that, although highly superheated droplets boil rapidly by the standards of macroscopic processes, there is nothing explosion-like about the process. The temperature of the bubble is largely confined to the range $T_{\text{sat}} - T_\infty$. The pressure in the bubble does not rise above the saturation pressure of the droplet liquid at T_∞ and the pressure caused by the expanding droplet drops off rapidly with distance from the bubble. There is nothing in the analysis that indicates that any significant shock wave would be caused by the boiling process.

References

Battya P, Raghavan VR, Seetharamu KN (1984) Parametric studies on direct contact evaporation of a drop in an immiscible liquid. Int J Heat Mass Transf 27:263–272

Brennen CE (1995) Cavitation and bubble dynamics. Oxford University Press, New York

Frost D, Sturtevant B (1986) Effects of ambient pressure on the instability of a liquid boiling explosively at the superheat limit. J Heat Transf Trans ASME 108:418–424

Kwak HY, Oh SD, Park CH (1995) Bubble dynamics on the evolving bubble formed from the droplet at the superheat limit. Int J Heat Mass Transf 38:1709–1718

Lee HS, Merte H (1996) Spherical vapor bubble growth in uniformly superheated liquids. Int J Heat Mass Transf 39:2427–2447

Lee HS, Merte H (2005) Explosive vapor bubble growth in uniformly superheated liquids: R-113 and mercury. Int J Heat Mass Transf 48:2593–2600

Lien Y-C (1969) Bubble growth rates at reduced pressure. Ph.D. Thesis, Massachusetts Institute of Technology, Cambridge, MA

Mahood HB (2008) Direct-contact heat transfer of a single volatile liquid drop evaporation in an immiscible liquid. Desalination 222:656–665

Mikic BB, Rohsenow WM, Griffith P (1970) On bubble growth rates. Int J Heat Mass Transf 13:657–666

Mori YH, Inui E, Komotori K (1978) Pool boiling heat transfer to emulsions. J Heat Transf Trans ASME 100:613–617

Park HC, Byun KT, Kwak HY (2005) Explosive boiling of liquid droplets at their superheat limits. Chem Eng Sci 60:1809–1821

Plesset MS, Prosperetti A (1977) Bubble dynamics and cavitation. Annu Rev Fluid Mech 9:148–185

Plesset MS, Zwick SA (1954) The growth of vapor bubbles in superheated liquids. J Appl Phys 25:493–500

Raina GK, Grover PD (1985) Direct contact heat transfer with change of phase: theoretical model incorporating sloshing effects. AIChE J 31:507–510

Rayleigh (1917) On the pressure developed in a liquid during the collapse of a spherical cavity. Philos Mag 34 (sixth series):94–98

Roesle ML, Kulacki FA (2010) Boiling of small droplets. Int J Heat Mass Transf 53:5587–5595

Shepherd JE, Sturtevant B (1982) Rapid evaporation at the superheat limit. J Fluid Mech 121:379–402

Sideman S, Taitel Y (1964) Direct-contact heat transfer with change of phase: evaporation of drops in an immiscible liquid medium. Int J Heat Mass Transf 7:1273–1289

Theofanous TG, Patel PD (1976) Universal relations for bubble growth. Int J Heat Mass Transf 19:425–429

Tochitani Y, Mori YH, Komotori K (1977a) Vaporization of single liquid drops in an immiscible liquid part 1: forms and motions of vaporizing drops. Heat Mass Transf 10:51–59

Tochitani Y, Nakagawa T, Mori YH et al (1977b) Vaporization of single liquid drops in an immiscible liquid part 2: heat transfer characteristics. Heat Mass Transf 10:71–79

Vuong ST, Sadhal SS (1989a) Growth and translation of a liquid-vapour compound drop in a second liquid, part 1: fluid mechanics. J Fluid Mech 209:617–637

Vuong ST, Sadhal SS (1989b) Growth and translation of a liquid-vapour compound drop in a second liquid, part 2: heat transfer. J Fluid Mech 209:639–660

Wohak MG, Beer H (1998) Numerical simulation of direct-contact evaporation of a drop rising in a hot, less volatile immiscible liquid of higher density—possibilities and limits of the SOLA-VOF/CSF algorithm. Numer Heat Transf Part A Appl 33:561–582

Chapter 4
A Model of Boiling in Emulsions

Keywords Eulerian multiphase model · Multiphase flow · Chain boiling · Bubble-droplet interaction · Effective property · Finite volume · OpenFOAM · PISO algorithm · CFD · Drag model

Boiling in boiling dilute emulsions requires analysis of heat and mass transport at multiple scales and of the behavior of the boiling droplet. Recall that the experimental work of Bulanov and co-workers shows that a boiling mechanism exists in addition to spontaneous nucleation. Also, the effects of the boiling droplets on the overall flow and heat transfer in the emulsion must be modeled.

In this chapter, a detailed model of boiling emulsions is developed based on observable phenomena. Interactions between superheated droplets and their surroundings that may cause boiling are characterized. This information on the behavior of individual droplets is used to define the parameters of a model of boiling emulsions based on the Euler–Euler approach introduced in Chap. 2. The primary goal in the development of this model is to account for all of the effects of boiling droplets in a dilute emulsion in a complete and physically consistent manner. In general, treatment of transport properties and closure equations is kept as simple as possible to avoid unnecessary complications at this stage of development. This model is then used in numerical simulations of boiling emulsions, which we have partially calibrated by experiment. A summary of the model development is given by Roesle and Kulacki (2010).

4.1 Balance Equations

A model of boiling emulsions can be developed using the Euler–Euler approach. The degree of superheat required for boiling may vary widely (Bulanov et al. 2006), therefore, the phase of the dispersed component cannot be easily determined based on either temperature or other macroscopic properties of the mixture. Thus, the liquid droplets and vapor bubbles must be treated as separate phases. The boiling emulsion is then modeled as a three-phase flow, where boiling and condensation of

M. L. Roesle and F. A. Kulacki, *Boiling Heat Transfer in Dilute Emulsions*, SpringerBriefs in Thermal Engineering and Applied Science, DOI: 10.1007/978-1-4614-4621-7_4, © The Author(s) 2013

the dispersed component results in mass transfer between the droplet and the bubble (vapor) phases. This approach also makes it possible to tailor constitutive relations to the droplet and bubble phases individually for quantities such as viscosity or drag.

Conservation of mass is applied separately to each phase. The boiling of liquid droplets is modeled as transfer of mass from the droplet phase to the vapor bubble phase, denoted $\dot{m}$. The droplet and carrier liquids are immiscible, so there is no mass transfer between them. It is assumed that the temperature of the emulsion remains below the saturation temperature of the continuous liquid, so that evaporation of the carrier liquid into the bubbles is negligible. The Boussinesq approximation for buoyancy is applied, so the density for each phase is assumed to be constant except for the buoyancy term in the momentum equations. Relations for the rate of boiling are developed in the following sections, based on the detailed behavior of the liquid and vapor particles. Starting with Eq. (2.3), the mass balance for each phase is,

$$\frac{\partial \varepsilon_c}{\partial t} + \nabla \cdot (\varepsilon_c \mathbf{U}_c) = 0, \tag{4.1}$$

$$\frac{\partial \varepsilon_b}{\partial t} + \nabla \cdot (\varepsilon_b \mathbf{U}_b) = +\frac{\dot{m}}{\rho_b}, \tag{4.2}$$

$$\frac{\partial \varepsilon_d}{\partial t} + \nabla \cdot (\varepsilon_d \mathbf{U}_d) = -\frac{\dot{m}}{\rho_d}. \tag{4.3}$$

Linear momentum is conserved for each phase. The only body force is gravity, and the momentum equation for each phase is obtained by applying the Boussinesq approximation to Eq. (2.4) and expressing the stress tensor as in Eq. (2.7). Expressions for the interfacial force terms are developed below. The notation $\mathbf{F}_{ij}$ indicates the average force exerted on phase i by phase j on a per-volume basis. Although the flow is assumed to be laminar in general, an expression for $\mu_{c,T}$ based upon the agitation of the flow by the boiling droplets is developed in the following sections as well. It is assumed that the pressure is the same in each phase.

It is also assumed that the droplet and bubble phases do not exert forces on each other directly. This assumption results from the fact that bubbles exist only within the thermal boundary layer. Outside the boundary layer, the bubbles quickly condense in the subcooled emulsion. Within the boundary layer, the droplets are superheated, so that when a droplet and bubble collide, the droplet boils. Thus, collisions between bubbles and droplets result in mass transfer between the two dispersed phases, which is more important than momentum transfer. Under these assumptions the phase momentum balances are,

$$\frac{\partial}{\partial t}(\varepsilon_c \mathbf{U}_c) + \nabla \cdot (\varepsilon_c \mathbf{U}_c \mathbf{U}_c) = \nabla \cdot \left[\frac{\varepsilon_c}{\rho_c}(\bar{\mu}_c + \mu_{c,T})(\nabla \mathbf{U}_c + \nabla^T \mathbf{U}_c - \frac{2}{3}(\nabla \cdot \mathbf{U}_c)\mathbf{I} \right]$$
$$- \frac{\varepsilon_c}{\rho_c}\nabla P + \varepsilon_c \mathbf{g}[1 - \beta_c(T - T_{c,0})] + \frac{\mathbf{F}_{cb}}{\rho_c} + \frac{\mathbf{F}_{cd}}{\rho_c}, \tag{4.4}$$

$$\frac{\partial}{\partial t}(\varepsilon_b \mathbf{U}_b) + \nabla \cdot (\varepsilon_b \mathbf{U}_b \mathbf{U}_b) = \nabla \cdot \left[\frac{\varepsilon_b \overline{\mu}_b}{\rho_b}\left(\nabla \mathbf{U}_b + \nabla^T \mathbf{U}_b - \frac{2}{3}(\nabla \cdot \mathbf{U}_b)\mathbf{I}\right)\right]$$
$$-\frac{\varepsilon_b}{\rho_b}\nabla P + \varepsilon_b \mathbf{g}\left[1 - \beta_b(T - T_{b,0})\right] + \frac{\dot{m}\mathbf{U}_b}{\rho_b} + \frac{\mathbf{F}_{bc}}{\rho_b},$$

$$(4.5)$$

$$\frac{\partial}{\partial t}(\varepsilon_d \mathbf{U}_d) + \nabla \cdot (\varepsilon_d \mathbf{U}_d \mathbf{U}_d) = \nabla \cdot \left[\frac{\varepsilon_d \overline{\mu}_d}{\rho_d}\left(\nabla \mathbf{U}_d + \nabla^T \mathbf{U}_d - \frac{2}{3}(\nabla \cdot \mathbf{U}_d)\mathbf{I}\right)\right]$$
$$-\frac{\varepsilon_d}{\rho_d}\nabla P + \varepsilon_d \mathbf{g}\left[1 - \beta_d(T - T_{d,0})\right] - \frac{\dot{m}\mathbf{U}_d}{\rho_d} + \frac{\mathbf{F}_{dc}}{\rho_d}.$$

$$(4.6)$$

The large degree of superheat required for boiling of dilute emulsions suggests that large temperature differences will be present in flows of boiling dilute emulsions. If it is assumed that fluid velocities are not very large, the contribution of kinetic energy to the conservation of energy equation becomes negligible. Further, because the droplets and bubbles are very small, the phases will be close to thermal equilibrium. The largest temperature variation would occur within a droplet when there is no circulation of fluid inside it so that heat transfer in the droplet is by pure conduction. For a quiescent droplet subjected to surroundings with steadily-rising temperature, the temperature difference between the droplet center and the surroundings is $(dT_\infty/dt)R_d^2/(6\alpha_c)$. For the size of droplets in emulsions and moderate rates of temperature change associated with droplets flowing into thermal boundary layers, the temperature variation inside droplets is negligible.

The internal energy equation can therefore be used, rather than the full energy equation, and only one mixture energy equation is necessary, rather than one energy equation for each phase. This approach also eliminates the need to model heat transfer between each phase and is a significant simplification (Bouré and Delhaye 1982). Starting with Eq. (2.5), the dissipation and heat source terms are neglected, and the internal energy is expressed in terms of temperature (assuming constant specific heat for each phase). The phase internal energy equations are then summed,

$$\frac{\partial}{\partial t}\left(T\sum_i \varepsilon_i \rho_i c_{v,i}\right) + \nabla \cdot \left(T\sum_i \varepsilon_i \rho_i c_{v,i}\mathbf{U}_i\right)$$
$$= \nabla \cdot \left[(k_{\text{eff}} + k_T)\nabla T\right] + \sum_i E_i + \sum_i c_{v,i} T_I \Gamma_i. \qquad (4.7)$$

The summations in Eq. (4.7) represent sums over all the phases in the emulsion. The symbol k_{eff} is the effective thermal conductivity of the emulsion and k_T is the turbulent thermal conductivity. The fourth term, the sum of the interfacial heat transfer into each phase, does not sum to zero due to phase change of the dispersed component. Consider for example the case of a droplet boiling as described in Chap. 3.

There is heat conduction in the droplet to the droplet–bubble interface (E_d is negative), but there is no heat conduction into the bubble. Instead, evaporation occurs ($\dot{m}$ is positive and equal to $-E_d/i_{fg}$). In the absence of phase change, the interfacial energy transfer terms sum to zero. The fourth term may therefore be rewritten in terms of the boiling rate, $-\dot{m}i_{fg}$. The last term in Eq. (4.7) may be eliminated by rewriting the first two terms of the equation,

$$\frac{\partial T}{\partial t}\sum_i \varepsilon_i\rho_i c_{v,i} + \sum_i \varepsilon_i\rho_i c_{v,i}\mathbf{U}_i\cdot\nabla T + T\sum_i c_{v,i}\rho_i\left[\frac{\partial\varepsilon_i}{\partial t} + \nabla\cdot(\varepsilon_i\mathbf{U}_i)\right]$$
$$= \nabla\cdot[(k_{\mathrm{eff}}+k_T)\nabla T] - \dot{m}i_{fg} + \sum_i c_{v,i}T_I\Gamma_i. \tag{4.8}$$

According to the phase continuity equation (Eq. 2.3), the expression in parentheses in the third term of Eq. (4.8) is equal to the mass transfer rate, Γ_i. It has already been assumed that the temperature is the same in each phase. It is reasonable to assume also that the temperature at the interface between phases is equal to the average temperature within the phases, so that the third and sixth terms of Eq. (4.8) cancel. The resulting mixture internal energy equation is,

$$\frac{\partial T}{\partial t}\sum_i \varepsilon_i\rho_i c_{v,i} + \sum_i \varepsilon_i\rho_i c_{v,i}\mathbf{U}_i\cdot\nabla T = \nabla\cdot[(k_{\mathrm{eff}}+k_T)\nabla T] - \dot{m}i_{fg}. \tag{4.9}$$

The above seven equations (Eqs. 4.1–4.6, 4.9) contain eight unknowns: ε_b, ε_c, ε_d, $\mathbf{U}_b$, $\mathbf{U}_c$, $\mathbf{U}_d$, T, and P. The requirement that the volume fractions sum to one may be used to eliminate one fraction as an independent variable, to bring the numbers of equations and unknowns into agreement. Several equations are also required to close this system of equations. Closure equations are needed to define fluid properties for each phase, as well as for the forces and mass transfer between the phases.

4.2   Closure Equations: Momentum, Mass, and Energy

4.2.1   Momentum Transfer

The forces between phases in Eqs. (4.4)–(4.6) must be defined. By Newton's third law of motion, $\mathbf{F}_{cd} = -\mathbf{F}_{dc}$ and $\mathbf{F}_{cb} = -\mathbf{F}_{bc}$, so only two sets of interfacial forces are needed. The droplets and bubbles in emulsions are sufficiently small so that the flow around them is Stokes' flow. Equation (2.8) is therefore used to define the drag force on each phase.

The small size of the droplets and bubbles in an emulsion also means that their drift velocity relative to that of the carrier fluid is very small. The expressions for virtual mass, lift, and rotational force (Eqs. 2.9–2.11) indicate that all three forces are proportional to the difference in velocity between the dispersed and continuous phases. These forces are negligible because the velocity difference is very small for emulsions. Turbulent drag (Eq. 2.12) is also neglected, both because it is proportional to the velocity difference between continuous and dispersed phases and because the flow is assumed to be laminar. Therefore, the interfacial force terms are,

$$\mathbf{F}_{dc} = -\mathbf{F}_{cd} = \frac{-18\varepsilon_d \mu_{\mathrm{eff}}}{d_d^2}(\mathbf{U}_d - \mathbf{U}_c), \tag{4.10}$$

$$\mathbf{F}_{bc} = -\mathbf{F}_{cb} = \frac{-18\varepsilon_b \mu_{\mathrm{eff}}}{d_b^2}(\mathbf{U}_b - \mathbf{U}_c). \tag{4.11}$$

4.2.2 Mass Transfer in Chain Boiling

Mass transfer occurs between the bubble and droplet phases as a result of boiling and condensation. Two causes of boiling are considered: contact between droplets and a heated surface and, collisions between droplets and bubbles. It will be shown that spontaneous nucleation is not a significant cause of boiling in dilute emulsions. Chain boiling will be considered as well, where boiling droplets collide with other nearby droplets. Condensation of bubbles is considered in areas where the emulsion is subcooled.

Chain boiling can occur when one boiling droplet causes adjacent droplets to boil as well. Bulanov and Gasanov (2008) discussed chain boiling and attributed it to the production of shock waves by boiling droplets. However, as is discussed in Chap. 3, there is no indication that a shock forms when a highly superheated droplet boils. Another possible mechanism for chain boiling is simple contact between the boiling droplet and an adjacent droplet.

The probability of a liquid droplet being close enough to a boiling droplet to make contact depends on the maximum diameter achieved by the boiling droplet, the local volume fraction of liquid droplets, and the motion of the adjacent droplets. The inertial response time due to Stokes drag of a particle (Peskin 1982) is,

$$\tau_{\mathrm{inertial}} = \frac{2}{9}\frac{R_d^2 \rho_d}{\mu_c}. \tag{4.12}$$

Based on this relation, the inertial response time of a typical droplet ($d_d \sim 10\ \mu\mathrm{m}$) in water is greater than the time required for a droplet to boil. Thus, droplets that surround a boiling droplet can be assumed to be stationary during the boiling process. Any droplet whose distance from a boiling droplet (measured center-to-center) is less than $R_d + R_{\mathrm{max}}$ will contact the boiling droplet, where R_{max} is the maximum radius achieved by the droplet during the boiling process.

This distance limit describes a sphere of volume $4/3\pi(R_d + R_{\mathrm{max}})^3$ centered on the boiling droplet. Clearly, if the number density of droplets in the emulsion is great enough that several droplets are contained in this volume on average, a chain reaction that causes most of the liquid droplets in the volume to boil rapidly can occur. Such chain reactions are probably responsible for the sudden foaming in the bulk reported by Mori et al. (1978) that brought some of their experiments to a sudden halt. That such a phenomenon is reported only by Mori et al. is most likely due to the fact that other researchers have not carried out their heat transfer experiments with the bulk temperature of the emulsion near the saturation temperature of the dispersed liquid. Because most experimenters examine heat transfer from a fine wire and use a bulk temperature significantly lower than the saturation temperature of the dispersed liquid, there never exists a large volume of super-heated emulsion. It is anticipated that under most circumstances the conditions necessary for such a chain reaction will not exist over a large volume.

Some chain boiling may occur at volume fractions lower than those necessary for the sustained chain reaction described above. The probability that a volume V does not contain a droplet when droplets are distributed randomly is $\exp(-N_d V)$. The probability of the volume containing at least one droplet is, therefore, $1-\exp(-N_d V)$. If it assumed that one and only one droplet is caused to boil when at least one droplet is less than the distance $R_d + R_{max}$ from a boiling droplet and that the distribution of droplets around a boiling droplet is independent of the number of droplets that have already boiled in the chain reaction, the probability of exactly φ droplets boiling in a chain reaction is $[1-\exp(-N_d V)]^{\varphi}\exp(-N_d V)$. The average number of droplets in such a chain reaction is then,

$$\overline{\varphi} = \sum_{\varphi=1}^{\infty} \varphi e^{-N_d V}\left(1 - e^{-N_d V}\right)^{\varphi} = \frac{\left(1 - e^{-N_d V}\right)}{e^{-N_d V}}. \tag{4.13}$$

The quantity $N_d V$ can be expressed in terms of the droplet volume fraction and the droplet and maximum bubble radii as,

$$N_d V = \varepsilon_d\left[\left(1+\frac{R_{max}}{R_d}\right)^3 -1\right]. \tag{4.14}$$

The rates of boiling by other causes described in the following sections should be multiplied by Eq. (4.13).

4.2.3 Mass Transfer in Boiling by Contact with Heated Surfaces

As discussed in Chap. 2, dilute emulsions can reach a very high degree of superheat before boiling occurs because most of the droplets suspended in the emulsion do not contact the heated surface. Because the temperature of the heated surface is significantly greater than the saturation temperature of the droplets, any droplets that may contact the surface will quickly boil. Whether and how many droplets will make contact depends on the flow geometry and conditions. It should be noted that this boiling mechanism is not limited to heated surfaces. If the emulsion is highly superheated, an insulated surface will have the same temperature as the emulsion and would, therefore, cause boiling.

Various forces may bring droplets into contact with wetted surfaces. For example, if the droplets are not neutrally buoyant they will tend to settle or rise and must eventually encounter the walls of the vessel containing the emulsion. In the case of flow in a vertical duct, non-neutrally buoyant droplets may migrate toward the walls due to lift forces (Haber and Hetsroni 1971). For droplets with sufficiently high Stokes numbers, the momentum of the droplet may carry it into contact with a wetted surface either when the flow changes direction suddenly (e.g., in the presence of an obstacle) or due to turbulent eddies (although here we generally limit analysis to laminar flow).

In most cases, the number of droplets contacting a heated solid surface will be very small compared to the total number of droplets, but this process will provide a supply of vapor bubbles to induce nucleation in other superheated droplets. When the liquid droplets are denser than the carrier liquid, any force that brings a droplet into contact with a surface will act in the opposite direction on the resulting vapor bubble, directing it back into the body of the flow. In the Euler–Euler approach to modeling multiphase flows, this behavior can be modeled by changing the boundary condition at heated surfaces to set the mass flux of droplet phase into the surface equal to the mass flux of bubble phase departing the surface.

4.2.4 Mass Transfer in Boiling by Bubble-Droplet Collisions

After bubbles are introduced into the superheated portion of an emulsion, they provide a source for further nucleation. A superheated liquid droplet will boil when it contacts a liquid–vapor interface such as a vapor bubble. Such contact can occur when bubbles and droplets collide owing to their relative motion. The relative motion may arise from shearing in the carrier liquid or from differing drift velocities of the droplets and bubbles.

Determining the rate of collisions between bubbles and droplets is complicated by the fact that they do not travel on straight lines when passing close to each other. The very small droplets and bubbles found in emulsions tend to follow the streamlines of the surrounding liquid. Therefore, a droplet and bubble that would be expected to collide based on their trajectories as they approach each other may instead move around each other without touching. If this behavior is neglected, calculation of the collision rate is straightforward. In the rest frame of a droplet, nearby bubbles move with velocity $\mathbf{U}_b - \mathbf{U}_d$. The bubble and droplet are represented by points located at their centers. As the bubble moves, over a time period Δt it sweeps out a cylindrical volume with radius $R_b + R_d$ and height $|\mathbf{U}_b - \mathbf{U}_d|\Delta t$. The bubble will collide with any droplet whose center is within this volume. In a mixture with many bubbles and many droplets, the collision rate for straight-line motion is,

$$J_{\text{coll},0} = \frac{9\varepsilon_d\varepsilon_b|\mathbf{U}_d - \mathbf{U}_b|(R_d + R_b)^2}{16\pi R_d^3 R_b^3} \tag{4.15}$$

The ratio between the number of actual collisions and the number of potential collisions predicted by Eq. (4.15) is the collision efficiency. The collision efficiency between two droplets has been studied for several decades in the context of raindrop formation in clouds, but studies of collision efficiency of bubbles and droplets suspended in liquid are not found in the literature. However, for very small droplets the Reynolds number of the flow around the droplet during steady rising or settling is close to zero so that the Stokes approximation is valid for the flow around the droplets. As long as this condition holds and if the droplets are treated as solid spheres, the collision efficiency for two particles depends only on their relative size, and not on the properties of the surrounding fluid or the absolute scale of the particles. From the data of Pinsky et al. (2001) for droplets in clouds that fall within the Stokes flow regime ($d < 20$ µm (Beard 1976)), for $4 \leq R_b/R_d \leq 6$ the collision efficiency, η, is quite small, decreasing from 0.01 to 0.006. Pinsky et al. treat their droplets as solid spheres, neglecting the effects of internal circulation. More accurate collision efficiencies could be calculated by considering the viscosity ratio between the particle and the surrounding fluid, but such data are not available.

Shearing of the continuous liquid will also cause particles in the flow to slide past one another. Given the small particle diameters found in emulsions and the very small collision efficiencies of particles in Stokes flow, only extremely strong shearing motion would produce a significant number of collisions. This source of collisions may need to be considered in emulsions with larger particles or flows of moderate velocity through small-diameter channels. In this study, however, this source of collisions will be neglected. In turbulent flows, the relative effects of turbulent eddies on bubbles and droplets must also be considered carefully. Here we consider only laminar flow so that these effects can be neglected.

To provide an equation for mass transfer rate due to this mechanism of boiling for use in the balance equations, the potential collision rate is multiplied by the collision efficiency, the average chain boiling length, and the mass of a droplet,

$$\dot{m}_{\text{coll}} = \varepsilon_b\varepsilon_d\rho_d|\mathbf{U}_d - \mathbf{U}_b|\eta\overline{\varphi}\frac{3(R_d + R_b)^2}{4R_b^3}. \tag{4.16}$$

4.2.5 Mass Transfer in Spontaneous Nucleation

A sufficiently superheated liquid will undergo spontaneous nucleation triggered by density fluctuations in the liquid. Equation (2.2) is easily adapted to predicting the nucleation rate of superheated droplets by multiplying it by the volume fraction of the droplet phase. However, the experimental studies by Bulanov et al. show that boiling in a dilute emulsion occurs over a wide temperature range starting at temperatures far lower than the kinetic limit of superheat. In contrast, the rate of spontaneous nucleation changes very rapidly, 10^2–10^3 K^{-1} for most liquids (Table 2.1). Thus, boiling due to spontaneous nucleation is negligible except in cases where the heated surface reaches the kinetic limit of superheat. Such cases are outside the scope of the present treatment.

Pressure fluctuations caused by a boiling droplet should be considered as well. When a droplet first begins to boil it accelerates the surrounding liquid outwards, thus increasing the pressure in the nearby liquid. The increased pressure has the effect of momentarily suppressing spontaneous nucleation in adjacent droplets, by decreasing the pressure difference in the denominator of the exponential factor in Eq. (2.2). On the other hand, when the bubble enters its final deceleration stage the pressure in the surrounding liquid is decreased and the probability of spontaneous nucleation in adjacent droplets rises. As the bubble oscillates around its equilibrium diameter the pressure in the surrounding liquid also oscillates, thus providing more time periods in which there is increased probability of spontaneous nucleation. However, these oscillations are quickly damped.

The effect of pressure oscillations is also small. The pressure inside a boiling droplet remains positive, so the maximum possible decrease in pressure in the surrounding liquid is less than the magnitude of the ambient pressure. Near the kinetic limit of superheat, the saturation pressure of fluids changes rapidly (Blander and Katz 1975). In an emulsion near atmospheric pressure, the momentary decrease in pressure caused by a boiling droplet has an effect equivalent to raising the emulsion temperature by only a few degrees. As noted above, emulsions begin to boil at temperatures far lower than the kinetic limit of superheat, so this effect is also negligible. It is possible that this effect may become significant for emulsions at high pressures.

4.2.6 Mass Transfer in Condensation

For emulsions in which the bulk temperature is lower than the saturation temperature of the dispersed fluid, condensation of bubbles will occur when they move out of the thermal boundary layer. Nucleation is not a concern here because the surface of the bubble provides a liquid–vapor interface upon which condensation can occur. The rate of condensation will be limited by heat transfer from the bubble. A rough estimate of the rate of condensation may be obtained based upon heat transfer from a sphere of radius R_b. For a sphere in quiescent fluid, the conduction limit gives $Nu_D = 2$, or $h = k_c/R_b$. Under these conditions the heat transfer from a bubble at its saturation temperature is,

$$q = 4\pi k_c R_b (T_{sat} - T). \qquad (4.17)$$

The condensation rate is obtained by dividing Eq. (4.17) by the latent heat of vaporization of the dispersed fluid. If it is assumed that the heat transfer remains

constant during the collapse of the bubble, the mass transfer due to condensation
for the emulsion is obtained by multiplying it by the bubble number density,

$$\dot{m}_{\text{cond}} = \min\left[-3\varepsilon_b \frac{k_c}{i_{\text{fg}} R_b^2}(T_{\text{sat}} - T), 0\right]. \tag{4.18}$$

Recall that $\dot{m}$ is defined as positive for boiling in the balance equations, and
therefore the mass transfer rate due to condensation must be negative. This
analysis gives only a rough estimate because it neglects the shell of droplet liquid
that grows around the bubble as it collapses. For emulsions in which the two
components have very different thermal conductivities, e.g., water and FC-72, the
impact of the droplet liquid is expected to be large. The condensation process is
not as critical as the boiling process, however, so this approximation is adequate.

4.2.7 Pseudo-Turbulent Effects: μ_T and k_T

The relations for mass and momentum transfer between the phases account for the
averaged behavior of the droplets and bubbles in the emulsion. This approach
neglects one potentially important effect of boiling droplets. Some of the improved
heat transfer that is observed in boiling emulsions has been attributed to agitation
of the emulsion caused by boiling droplets (Bulanov et al. 1996; Ostrovskiy 1988).
The effects of this agitation can be modeled in an Euler–Euler model by borrowing
a technique from turbulence modeling.

In many turbulent flows, the most significant effect of the small-scale turbulent
eddies on the averaged large-scale behavior of the flow is an increase in the rate of
momentum transport in the direction of the mean velocity gradient in the flow. This
effect occurs because the turbulent eddies move elements of fast moving fluid into
regions of slow moving fluid and vice versa. The effect can be modeled by introducing a turbulent kinematic viscosity, v_T, in the equations used to model the mean
flow. Similarly, the turbulent eddies also increase heat transfer in the direction of the
mean temperature gradient, which can be modeled by introducing a turbulent
thermal diffusivity, α_T. Experiments show that for fluids with $Pr \sim 1$, the turbulent
Prandtl number, $Pr_T = \alpha_T / v_T$, is ~ 0.85 (Pope 2000; Kays et al. 2005). It is therefore
only necessary to model the turbulent viscosity, and then the turbulent thermal
diffusivity can be obtained using the Reynolds analogy.

One approach to modeling turbulent viscosity is to express it as the product of a
characteristic length l and a characteristic velocity u^*,

$$v_T = u^* l. \tag{4.19}$$

The characteristic length represents the size of a turbulent eddy, and the
characteristic velocity can be thought of as the average eddy velocity. In turbulent
flow, determining u^* and l is a subtle and complex problem (Pope 2000). On the
other hand, in boiling emulsions both quantities can be related directly to the
properties of the boiling droplets.

When a droplet boils, it displaces the liquid around it symmetrically outward. If the volume of the initial droplet is small compared to the volume of the bubble, the displacement, Δr, of a fluid element initially located a distance r_{init} from the center of the droplet is,

$$\Delta r = \left(r_{\text{init}}^3 + R_b^3\right)^{1/3} - r_{\text{init}}. \tag{4.20}$$

The magnitude of the displacement, therefore, decreases quickly with distance from the boiling droplet. At a distance of twice the bubble radius, the displacement is less than $0.1\ R_b$, so it is reasonable to use this distance as a cutoff as the displacement due to boiling of liquid outside of this radius is insignificant. The average displacement of the liquid in this volume is obtained by integrating Eq. (4.20) over the volume, and the result is $\approx 0.16\ R_b$. This displacement is directed radially from the droplet, but for this pseudo-turbulent model, only the component of the displacement perpendicular to the direction of a mean gradient in the flow is important. The average displacement in one dimension is half the radial displacement, so $l \approx 0.08\ R_b$.

The characteristic velocity can be thought of as the volume-averaged velocity of the liquid being displaced. A value for u^* can be expressed as the product of the volumetric rate of droplet nucleation, the volume of displaced liquid per boiling droplet, and the average displacement of the liquid,

$$u^* = J(0.16R_b)\left[\frac{4}{3}\pi(2R_b)^3\right] = 1.28\,JV_bR_b. \tag{4.21}$$

By substituting these values for u^* and l into Eq. (4.19), the turbulent viscosity is

$$v_T = 0.1\,JV_bR_b^2. \tag{4.22}$$

For use in the balance equations, the nucleation rate in Eq. (4.22) can be expressed in terms of the rate of mass transfer between the droplet and bubble phases,

$$\mu_{c,T} = 0.1\dot{m}\frac{\rho_c}{\rho_b}R_b^2. \tag{4.23}$$

As discussed above, the Reynolds analogy may be used to obtain a turbulent thermal conductivity from Eq. (4.23),

$$k_T = 0.1\dot{m}c_{p,c}\frac{\rho_c}{\rho_b}R_b^2. \tag{4.24}$$

This model accounts for the net outward displacement of the continuous liquid when isolated droplets boil. However, when a single droplet boils, the bubble initially grows to a radius significantly larger than its equilibrium size and then oscillates for some time through radial expansion and contraction (Chap. 3). These repeated displacements of the surrounding liquid may result in a much larger effect than the single outward displacement modeled here. The precise

effects of the oscillating bubble on the surrounding liquid are not known, especially when several droplets boil in proximity to each other or in regions where
there is strong shearing of the continuous liquid. If the agitation of the surrounding
liquid is not equal to Eqs. (4.23) and (4.24), it is reasonable to model the effects as
at least being proportional to these equations.

4.2.8 Effective Viscosity

Equations (4.4)–(4.6) require an effective viscosity for each phase, which is generally not equal to the viscosity of the fluid that constitutes that phase. As described
in Chap. 2, there does not appear to be any consensus as to how this problem should
be approached. Here we adopt a two step process, in which an effective viscosity for
the mixture is calculated and then is distributed to each of the phases.

It is worthwhile to start by considering the desired asymptotic behavior of the
correlations for effective viscosity. First, when only one phase is present the
effective viscosity should be equal to that of the fluid, that is, $\bar{\mu}_i \to \mu_i$ as $\varepsilon_i \to 1$. It
is important that each phase exhibit this behavior because, even though the
emulsions to be studied are dilute, the volume fraction of the bubble phase may be
much higher than the average volume fraction of the droplet phase where boiling
occurs. The volume fraction of the droplet phase may also become large locally in
cases such as droplets settling onto a horizontal surface. At the other extreme, the
effective viscosity of a phase should go to zero as its volume fraction approaches
zero. Recall that the effective viscosity of a dispersed phase represents its ability to
transmit shear stress through interactions between elements of it independently of
interactions between the dispersed and the continuous phases. When the volume
fraction of the dispersed phase is small, the elements of the dispersed phase are
widely separated and should have very little direct interaction with each other.

None of the strategies for assigning effective viscosity to each phase described
in Chap. 2 exhibit these asymptotic behaviors. In the absence of theoretical
arguments for any particular weighting of viscosities between the phases, the
mixture viscosity is weighted by the local volume fraction of each phase,

$$\bar{\mu}_i = \varepsilon_i \mu_{\text{eff}}. \tag{4.25}$$

A correlation for the effective viscosity of the mixture is required. The correlation should be as simple as possible so that it can be used in numerical simulations, and it must be valid for a wide range of phase fractions of the dispersed
phase. The latter requirement rules out most theoretical correlations, such as
developed by Taylor (1932), because they are generally valid only for dilute
mixtures. Many empirical correlations exist for the effective viscosity of liquid–
vapor mixtures under various conditions. A correlation that has been shown
experimentally to be valid for bubbly flow and that has the correct asymptotic
behavior for dilute mixtures is that of Beattie and Whalley (1982), which is
expanded to include two dispersed phases,

$$\mu_{\text{eff}} = \mu_c \varepsilon_c [1 + 2.5(\varepsilon_b + \varepsilon_d)] + \mu_b \varepsilon_b + \mu_d \varepsilon_d. \tag{4.26}$$

When Eqs. (4.25) and (4.26) are used to assign viscosities to each phase, the assigned viscosities have reasonable asymptotic behavior, in that that $\overline{\mu}_i \to 0$ as $\varepsilon_i \to 0$, $\overline{\mu}_i \to \mu_i$ as $\varepsilon_i \to 1$, and $\mu_{\text{eff}} \to \mu_i$ as $\varepsilon_i \to 1$ (and $\varepsilon_j \to 0$, $j \neq i$).

4.2.9 Effective Thermal Conductivity

The effective thermal conductivity of the mixture must also be defined. The use of a single energy equation for the mixture obviates assigning conductivities to each phase individually. A similar difficulty is encountered as in the determination of effective viscosities. Simple theoretical models exist, e.g., Maxwell's effective medium theory, but are applicable only when the volume fraction of the dispersed phase is small (Maxwell 1904). Theoretical and empirical correlations valid for higher fractions exist, but all are much more complicated. The more complex correlations generally fall close to the results of the effective medium theory even at moderate volume fractions (Buyevich 1992), so it is assumed here that the effective medium theory holds. It is further assumed that the effects of the bubble phase and the droplet phase are additive. A correlation for the mixture effective thermal conductivity is therefore,

$$k_{\text{eff}} = k_c + \frac{3\varepsilon_d(k_d - k_c)}{\varepsilon_d + \frac{\varepsilon_c k_d}{k_c} + 2} + \frac{3\varepsilon_b(k_b - k_c)}{\varepsilon_b + \frac{\varepsilon_c k_b}{k_c} + 2}. \tag{4.27}$$

4.3 Numerical Model and Solution

The model of boiling emulsions described in Sects. 4.1 and 4.2 must be implemented numerically. This section describes the numerical solution procedure, and solutions are presented in Chap. 6 in connection with heat transfer measurements.

The numerical approach borrows heavily from that of Rusche (2002), as adapted to multiple dispersed phases by Silva and Lage (2007). The solver we have used is based on the finite volume numerical method implemented using the OpenFOAM™ (Field Operation and Manipulation) computational fluid dynamics (CFD) package.[1] It provides a library of routines for manipulating volumetric fields and setting up and solving partial differential equations using the finite volume approach.

---

[1] OpenFOAM™ is a free CFD package written in C++ and distributed under the GNU general public license (GPL). Details of the operation of OpenFOAM™ and its numerical implementation of vector and differential operators may be found in the User Guide (2009) and Programmer's Guide (2009).

Coupling between pressure and velocity is accomplished with the pressure implicit with splitting of operators (PISO) algorithm. The PISO algorithm is a predictor–corrector type procedure in which at each time step, a prediction is made for the new velocity field, and then the new pressure field is solved such that continuity is enforced. The algorithm is similar to the SIMPLE algorithm, but uses multiple iterations of the corrector step and is suitable for non-stationary flows. The phase volume fraction, velocity, and temperature equations are linked non-linearly through mass transfer and buoyancy, so each time step consists of an outer loop in which the volume fraction, velocity, and temperature fields are solved iteratively as well as inner loops for the volume fraction and velocity fields.

It is necessary now to introduce nomenclature to describe how the equations developed in the previous chapter are discretized and solved. In the finite volume approach, the solution domain is split into a number of cells separated by faces. The dependent variables, $\mathbf{U}_i$, ε_i, and T, are considered cell-centered variables, meaning that they represent average values in each cell in the domain. In the absence of high order of accuracy numerical methods, these variables can also be thought of as representing the values at the centroid of each cell. Similarly, face-centered variables are defined for each face in the domain and represent the average value at each face. A cell-centered variable can be interpolated to the face centers, which is indicated with the subscript F. One important face-centered variable used in the following solution procedure is the phase volumetric flux ϕ_i. The phase volumetric flux is defined as the dot product of the cell face area vector and the phase velocity extrapolated to the cell faces, $\phi_i = \mathbf{S} \cdot (\mathbf{U}_i)_F$.

Averaging of a variable over several neighboring cells is denoted with angular brackets, with a subscript indicating the pattern of cells over which averaging takes place. For example, $\langle \varepsilon_i \rangle_\nabla$ is the average of ε_i over the same cells used for computing the gradient. This operation is important in dealing with the term $\nabla \varepsilon_i / \varepsilon_i$, which appears in the phase-intensive momentum equation below. In general, $\nabla \varepsilon_i$ should become small as $\varepsilon_i \to 0$. However, the case can arise where ε_i is zero in a cell but not in one of its neighbors. In such a case it is not possible to divide by the value of ε_i in the cell in question, but the average value $\langle \varepsilon_i \rangle_\nabla$ is guaranteed to be nonzero when $\nabla \varepsilon_i$ is nonzero (recognizing also that ε_i is non-negative). Rusche (2002) adds an additional stabilizing factor, $\nabla \varepsilon_i / (\langle \varepsilon_i \rangle_\nabla + \Delta)$.

The mass, momentum, and energy balances are treated implicitly in time, so the result of discretizing each of them is a system of algebraic equations that must be solved simultaneously. Implicit discretization is denoted $\| \mathscr{L}[x] \|$, where the operator $\mathscr{L}$ is discretized implicitly in terms of the variable x. (Terms without the double brackets are discretized explicitly). All terms in an equation that are treated implicitly must be discretized in terms of the same variable. Systems of equations resulting from implicit discretization are denoted with script variables. For example, the generic balance equation (Appendix A, Eq. A1) may be discretized implicitly in its first two terms as,

$$\mathscr{A} := \left\{ \left\| \frac{\partial \rho[\Psi]}{\partial t} \right\| + \left\| \nabla \cdot (\rho \mathbf{U}[\Psi]_F) \right\| = \nabla \cdot \gamma + \rho \xi \right\}. \tag{4.28}$$

Here, $\mathscr{A}$ is the system of linear algebraic equations representing the discretized balance equation. In this example, in the absence of a specific model for γ, the diffusion term must be discretized explicitly. A number of operators may now be defined to refer to specific features of the system of equations (Rusche 2002). Some important ones are $\mathscr{A}_D$ and $\mathscr{A}_N$ for the diagonal and off diagonal components of the matrix coefficients, respectively. The source vector is denoted by $\mathscr{A}_S$. Finally, the H operator is defined,

$$\mathscr{A}_H \equiv \mathscr{A}_S - \mathscr{A}_N \Psi, \tag{4.29}$$

where Ψ is the dependent variable of the discretized equations.

4.3.1 Momentum Equations

As the volume fraction of a phase becomes small, the phase momentum (Eqs. 4.4–4.6) become singular. The solution adopted by Rusche (2002) is to divide the momentum equation by the phase volume fraction, which gives the phase-intensive momentum equation. For brevity, it is helpful to briefly return to the generic phase momentum equation of Eq. (2.4) and represent the viscous stress term simply with $\mathbf{R}_{\text{eff},i}$, as in Eq. (2.7). The momentum equation is first divided by density (assumed constant) and the first three terms are expanded,

$$\mathbf{U}_i \left[\frac{\partial \varepsilon_i}{\partial t} + \nabla \cdot (\varepsilon_i \mathbf{U}_i) \right] + \varepsilon_i \left[\frac{\partial \mathbf{U}_i}{\partial t} + \mathbf{U}_i \cdot \nabla \mathbf{U}_i \right] + \frac{\varepsilon_i}{\rho_i} \nabla \cdot \mathbf{R}_{\text{eff},i} + \mathbf{R}_{\text{eff},i} \cdot \frac{\nabla \varepsilon_i}{\rho_i}$$
$$= -\varepsilon_i \nabla P + \varepsilon_i \mathbf{b}_i + \frac{\mathbf{F}_i}{\rho_i} + \frac{1}{\rho_i} \mathbf{U}_{i,I} \Gamma_i. \tag{4.30}$$

According to the phase mass balance (Eq. 2.3), the first term of Eq. (4.30) is equal to $\mathbf{U}_i \Gamma_i / \rho_i$. It is assumed that the average phase velocity, $\mathbf{U}_i$, is equal to the average velocity of fluid entering the phase across an interface, $\mathbf{U}_{i,I}$, so that the first term and last terms of Eq. (4.30) are equal and cancel out. Dividing by the phase volume fraction, one obtains the phase-intensive momentum equation,

$$\frac{\partial \mathbf{U}_i}{\partial t} + \mathbf{U}_i \cdot \nabla \mathbf{U}_i + \nabla \cdot \mathbf{R}_{\text{eff},i} + \mathbf{R}_{\text{eff},i} \cdot \frac{\nabla \varepsilon_i}{\varepsilon_i} = -\frac{\nabla P}{\rho_i} + \mathbf{g} \left[1 - \beta_i (T - T_{i,0}) \right] + \frac{\mathbf{F}_i}{\varepsilon_i \rho_i}.$$
$$\tag{4.31}$$

Here the generic body force $\mathbf{b}_i$ has been replaced with the gravitational force as in Eqs. (4.4)–(4.6). Equation (4.31) contains two terms that contain the phase volume fraction in the denominator. The fourth term should always be finite because $\nabla \varepsilon_i \to 0$ as $\varepsilon_i \to 0$, and the term is discretized as described above to avoid division by zero in the numerical algorithm. The final term does not present a problem for the dispersed phases because, according to Eqs. (4.10) and (4.11), the interfacial force term for each dispersed phase goes to zero as the volume fraction of the phase goes to zero. However if the volume fraction of the continuous phase can become small at any location during a simulation, careful treatment of the interfacial force terms for the continuous phase is required. A method for handling such a situation is described below.

For discretization of the momentum equations, Rusche (2002) splits the effective phase stress into a diffusive component and a correction,

$$\mathbf{R}^D_{\text{eff},i} = -\nu_{\text{eff},i} \nabla \mathbf{U}_i,$$

$$\mathbf{R}^C_{\text{eff},i} = -\nu_{\text{eff},i} \left[\nabla^T \mathbf{U}_i - \frac{2}{3}\mathbf{I}(\nabla \cdot \mathbf{U}_i)\right], \tag{4.32}$$

so that $\mathbf{R}^D_{\text{eff}} + \mathbf{R}^C_{\text{eff}} = \mathbf{R}_{\text{eff}}$, in accordance with the definition of the effective stress in Eq. (2.7). Rusche then discretizes the left hand side of Eq. (4.31) as,

$$\mathscr{F}_i = \left\|\frac{\partial [\mathbf{U}_i]}{\partial t}\right\| + \left\|\nabla \cdot (\phi_i[\mathbf{U}_i]_F)\right\| + \left\|\nabla \cdot \left(\nu_{iF}\frac{\nabla \varepsilon_i}{\varepsilon_{iF} + \Delta}[\mathbf{U}_i]_F\right)\right\| - \left\|[\mathbf{U}_i]\nabla \cdot \phi_i\right\|$$

$$+ \left\|[\mathbf{U}_i]\nabla \cdot \left(\nu_{iF}\frac{\nabla \varepsilon_i}{\varepsilon_{iF} + \Delta}\right)\right\| - \left\|\nabla \cdot (\nu_{iF}\nabla[\mathbf{U}_i])\right\| + \nabla \cdot \mathbf{R}^C_{\text{eff},i} + \frac{\nabla \varepsilon_i}{\langle \varepsilon_i \rangle_\nabla + \Delta} \cdot \mathbf{R}^C_{\text{eff},i} \tag{4.33}$$

Next, interfacial momentum transfer is considered. Rusche (2002) discusses methods for discretizing lift, drag, and virtual mass forces. In this study, only drag forces are considered and the same semi-implicit method is used. The drag force applied to phase i by phase j is,

$$\mathbf{F}_{ij} = \frac{18\varepsilon\mu_{\text{eff}}}{d^2}\left(\mathbf{U}_j - \|\mathbf{U}_i\|\right). \tag{4.34}$$

The phase volume fraction and diameter in Eq. (4.34) are those of the dispersed phase. The implicit portion of the drag force is combined with Eq. (4.33),

$$\mathscr{A}_b := \left\{\mathscr{F}_b = -\left\|\frac{18\mu_{\text{eff}}}{d_b^2\rho_b}[\mathbf{U}_b]\right\|\right\},$$

$$\mathscr{A}_c := \left\{\mathscr{F}_c = -\left\|\frac{18\varepsilon_b\mu_{\text{eff}}}{d_b^2\varepsilon_c\rho_c}[\mathbf{U}_c]\right\| - \left\|\frac{18\varepsilon_d\mu_{\text{eff}}}{d_d^2\varepsilon_c\rho_c}[\mathbf{U}_c]\right\|\right\}, \tag{4.35}$$

$$\mathscr{A}_d := \left\{\mathscr{F}_d = -\left\|\frac{18\mu_{\text{eff}}}{d_d^2\rho_d}[\mathbf{U}_d]\right\|\right\}.$$

The complete phase momentum equation can then be expressed in semi-discretized form as,

$$(\mathscr{A}_i)_D\mathbf{U}_i = (\mathscr{A}_i)_H - \frac{\nabla P}{\rho_i} + \mathbf{g}\left[1 - \beta_i\left(T - T_{i,0}\right)\right] + \frac{18\varepsilon\mu_{\text{eff}}}{d^2\varepsilon_i\rho_i}\mathbf{U}_j. \tag{4.36}$$

Equations (4.35) and (4.36) are equivalent to the phase momentum balance equations, Eqs. (4.4)–(4.6), with the addition of the momentum transfer closure equations, Eqs. (4.10) and (4.11).

Equation (4.36) is not solved directly. Instead, it is used to define predictor and corrector equations for the face volumetric flux, which are used in the PISO algorithm to obtain the velocity and pressure fields. The use of face flux as the primary variable in the PISO loop eliminates the need for a staggered grid to avoid checker boarding of pressure (Patankar 1980; Rusche 2002). First, Eq. (4.36) is solved for the phase velocity,

$$\mathbf{U}_i = \frac{(\mathscr{A}_i)_H}{(\mathscr{A}_i)_D} - \frac{\nabla P}{\rho_i(\mathscr{A}_i)_D} + \frac{\mathbf{g}[1 - \beta_i(T - T_{\mathrm{ref},i})]}{(\mathscr{A}_i)_D} + \frac{18\varepsilon\mu_{\mathrm{eff}}\mathbf{U}_j}{d^2\varepsilon_i\rho_i(\mathscr{A}_i)_D}. \qquad (4.37)$$

Recalling the definition of volumetric face flux, the equation for face flux can be obtained from Eq. (4.37) by interpolating each term to the cell faces and taking the dot product of each with the face area vector, $\mathbf{S}$. The left hand side of Eq. (4.37) is thus the phase volumetric flux. The first term on the right hand side is interpolated to the face centers and is not manipulated further. The flux predictor equation omits the pressure term, which will be reintroduced in the flux corrector equation. In the third term on the right hand side, the gravity vector (a constant) requires no interpolation. In the last term of Eq. (4.37), the phase velocity is also converted to a phase volumetric flux. Thus, the flux predictor equation for each phase is,

$$\phi_b^* = \left(\frac{(\mathscr{A}_b)_H}{(\mathscr{A}_b)_D}\right)_F \cdot \mathbf{S} + \left(\frac{1 - \beta_b(T - T_{\mathrm{ref},b})}{(\mathscr{A}_b)_D}\right)_F \mathbf{g} \cdot \mathbf{S} + \left(\frac{18\mu_{\mathrm{eff}}}{d_b^2\rho_b(\mathscr{A}_b)_D}\right)_F \phi_c,$$

$$\phi_c^* = \left(\frac{(\mathscr{A}_c)_H}{(\mathscr{A}_c)_D}\right)_F \cdot \mathbf{S} + \left(\frac{1 - \beta_c(T - T_{\mathrm{ref},c})}{(\mathscr{A}_c)_D}\right)_F \mathbf{g} \cdot \mathbf{S} + \left(\frac{18\varepsilon_b\mu_{\mathrm{eff}}}{d_b^2\varepsilon_c\rho_c(\mathscr{A}_c)_D}\right)_F \phi_b$$

$$+ \left(\frac{18\varepsilon_d\mu_{\mathrm{eff}}}{d_d^2\varepsilon_c\rho_c(\mathscr{A}_c)_D}\right)_F \phi_d,$$

$$\phi_d^* = \left(\frac{(\mathscr{A}_d)_H}{(\mathscr{A}_d)_D}\right)_F \cdot \mathbf{S} + \left(\frac{1 - \beta_d(T - T_{\mathrm{ref},d})}{(\mathscr{A}_d)_D}\right)_F \mathbf{g} \cdot \mathbf{S} + \left(\frac{18\mu_{\mathrm{eff}}}{d_d^2\rho_d(\mathscr{A}_d)_D}\right)_F \phi_c, \qquad (4.38)$$

The flux corrector equation has the same form for each phase,

$$\phi_i = \phi_i^* - \left(\frac{1}{\rho_i(\mathscr{A}_i)_D}\right)_F \nabla P \cdot \mathbf{S}. \qquad (4.39)$$

The pressure field is used to enforce continuity in the PISO algorithm, so the pressure equation stems from the continuity equation. The multiphase model used in this study assumes that the pressure is the same in each phase, so the mixture continuity equation is used, which is obtained by summing Eqs. (4.1)–(4.3),

$$\nabla \cdot (\varepsilon_b \mathbf{U}_b + \varepsilon_c \mathbf{U}_c + \varepsilon_d \mathbf{U}_d) = \dot{m}\left(\frac{1}{\rho_b} - \frac{1}{\rho_d}\right). \tag{4.40}$$

The left hand side of Eq. (4.40) is interpolated to the face centers and the velocities are expressed in terms of the face volume fluxes,

$$\nabla \cdot (\varepsilon_{bF} \phi_b + \varepsilon_{cF} \phi_c + \varepsilon_{dF} \phi_d) = \dot{m}\left(\frac{1}{\rho_b} - \frac{1}{\rho_d}\right). \tag{4.41}$$

Next, Eq. (4.39) is substituted in for each phase flux and the equation is rearranged to,

$$\left\| \nabla \cdot \left(\left(\frac{\varepsilon_{bF}}{\rho_b} \left(\frac{1}{(\mathscr{A}_b)_D} \right)_F + \frac{\varepsilon_{cF}}{\rho_c} \left(\frac{1}{(\mathscr{A}_c)_D} \right)_F + \frac{\varepsilon_{dF}}{\rho_d} \left(\frac{1}{(\mathscr{A}_d)_D} \right)_F \right) \nabla[P] \right) \right\|$$
$$= \nabla \cdot \left(\varepsilon_{bF} \phi_b^* + \varepsilon_{cF} \phi_c^* + \varepsilon_{dF} \phi_d^* \right) - \dot{m}\left(\frac{1}{\rho_b} - \frac{1}{\rho_d}\right). \tag{4.42}$$

This equation can be solved for the pressure field after the face fluxes are predicted using Eq. (4.38). Finally, the updated velocity field is reconstructed from the corrected face fluxes.

4.3.2 Phase Continuity Equations

The phase continuity equations (Eqs. 4.1–4.3) are solved to update the phase volume fractions at each time step. Rusche (2002) discusses several methods of solving the phase continuity equations for a two-phase mixture. He notes that the most important feature of the solution procedure is that it produces results that are bounded and conservative. The method preferred by Rusche is one that accounts for coupling between the phases using the relative velocity of the two phases, which also improves the efficiency of the solution. This method is expanded to mixtures with multiple dispersed phases by Silva and Lage (2007), and their method is used in this work. The essential features of the method are as follows.

First, the mixture velocity is defined as the volumetric average of the phase velocities,

$$\mathbf{U}_m = \sum_i \varepsilon_i \mathbf{U}_i. \tag{4.43}$$

and the relative velocity is defined as,

$$\mathbf{U}_{r,ij} = \mathbf{U}_i - \mathbf{U}_j. \tag{4.44}$$

The velocity of a phase can then be written as,

$$\mathbf{U}_i = \mathbf{U}_m + \sum_{j \neq i} \varepsilon_i \mathbf{U}_{r,ij}. \tag{4.45}$$

Substituting Eq. (4.45) into the incompressible version of the phase continuity equation (Eq. 2.3) yields,

$$\frac{\partial \varepsilon_i}{\partial t} + \nabla \cdot (\mathbf{U}_m \varepsilon_i) + \nabla \cdot \left(\sum_{j \neq i} \varepsilon_j \mathbf{U}_{r,ij} \varepsilon_i \right) = \frac{\Gamma_i}{\rho_i}. \tag{4.46}$$

This equation is discretized as,

$$\left\| \frac{\partial [\varepsilon_i]}{\partial t} \right\| + \left\| \nabla \cdot (\phi_m[\varepsilon_i]) \right\| + \left\| \nabla \cdot \left(\sum_{j \neq i} \varepsilon_j \phi_{r,ij} [\varepsilon_i] \right) \right\|$$

$$= \left\| \frac{\Gamma_{i,+}}{\rho_i} \left(\frac{[\varepsilon_i] - 1}{1 - \varepsilon_i^0 - \Delta} \right) \right\| + \left\| \frac{\Gamma_{i,-}}{\rho_i} \left(\frac{[\varepsilon_i]}{\varepsilon_i^0 + \Delta} \right) \right\|. \tag{4.47}$$

The mass transfer term in Eq. (4.47) is discretized as recommended by Patankar (1980) to ensure that the result remain bounded between 0 and 1.

In practice, Eq. (4.47) need only be solved for the dispersed phases,

$$\left\| \frac{\partial [\varepsilon_b]}{\partial t} \right\| + \left\| \nabla \cdot (\phi_m[\varepsilon_b]) \right\| + \left\| \nabla \cdot (\varepsilon_c \phi_{r,bc} [\varepsilon_b] + \varepsilon_d \phi_{r,bd} [\varepsilon_b]) \right\|$$

$$= \left\| \frac{\dot{m}_+}{\rho_b} \left(\frac{[\varepsilon_b] - 1}{1 - \varepsilon_b^0 - \Delta} \right) \right\| + \left\| \frac{\dot{m}_-}{\rho_b} \left(\frac{[\varepsilon_b]}{\varepsilon_b^0 + \Delta} \right) \right\|,$$

$$\left\| \frac{\partial [\varepsilon_d]}{\partial t} \right\| + \left\| \nabla \cdot (\phi_m[\varepsilon_d]) \right\| + \left\| \nabla \cdot (\varepsilon_b \phi_{r,db} [\varepsilon_d] + \varepsilon_c \phi_{r,dc} [\varepsilon_d]) \right\| \tag{4.48}$$

$$= \left\| \frac{-\dot{m}_-}{\rho_d} \left(\frac{[\varepsilon_b] - 1}{1 - \varepsilon_d^0 - \Delta} \right) \right\| + \left\| \frac{-\dot{m}_+}{\rho_d} \left(\frac{[\varepsilon_d]}{\varepsilon_d^0 + \Delta} \right) \right\|,$$

The volume fraction of the continuous phase is determined by the requirement that the volume fractions add to one at every point,

$$\varepsilon_c = 1 - \varepsilon_b - \varepsilon_d. \tag{4.49}$$

Equation (4.48) is nonlinear, so the set of phase continuity equations must be solved iteratively to achieve a converged solution.

4.3.3 Internal Energy and Phase Change Equations

The internal energy equation (Eq. 4.9) is discretized in a straightforward manner. The second term is expanded so that all spatial derivatives are expressed as divergences,

$$\left\| \frac{\partial [T]}{\partial t} \sum_i \varepsilon_i \rho_i c_{v,i} \right\| + \left\| \sum_i \rho_i c_{v,i} \nabla \cdot (\varepsilon_{iF} \phi_i [T]) \right\| - \left\| \sum_i \rho_i c_{v,i} [T] \nabla \cdot (\varepsilon_{iF} \phi_i) \right\| \qquad (4.50)$$
$$= \| \nabla \cdot ((k_{\text{eff}} + k_T) \nabla [T]) \| - \dot{m} i_{fg}.$$

Mass transfer is calculated using Eqs. (4.16) and (4.18). Equation (4.16) is applied only where $T > T_{\text{sat}}$ for the dispersed component and is limited to less than $\varepsilon_d^o \rho_d / \Delta t$, which is the boiling rate that would consume all of the droplets within a single time step. Similarly, the condensation rate is limited to $\varepsilon_b^o \rho_b / \Delta t$. Boiling at the heated surface is handled as described in Sect. 4.2. The convection and buoyant forces that would bring droplets into contact with the heated surface are already included in the numerical model. Accordingly, the rate at which droplets contact the heated surface is determined using the cell-centered droplet velocity and volume fraction in the cells adjacent to the heated surface. For each cell adjacent to the heated surface,

$$\dot{m}_w = \frac{\rho_d \varepsilon_d \mathbf{U}_d \cdot \mathbf{S}_w}{V}, \qquad (4.51)$$

where V is the volume of the cell. Thus, boiling due to wall contact is not implemented as a true boundary condition in the finite volume numerical method. Implementation of such a boundary condition would pose significant difficulties, since the boundary condition would couple the ε_b, ε_d, $\mathbf{U}_b$, and $\mathbf{U}_d$ fields.

4.3.4 Solution Procedure

The pressure, volume fraction, and temperature equations are coupled in a non-linear fashion by the boiling and condensation of the dispersed component. Thus, it is necessary to perform several iterations of the balance equations at each time step. Iterations are performed until the mass transfer rates reach convergence, as determined by a mixed absolute and relative tolerance similar to that used by Silva and Lage (2007),

$$\max \left[\frac{|\dot{m} - \dot{m}^o|}{1 \sim \kappa \gamma / \text{m}^3 \text{s} + |\dot{m}|} \right] < 0.01. \qquad (4.52)$$

The solution procedure for each time step is to first update the phase fractions by solving the dispersed phase volume fraction (Eq. 4.48) and calculating the continuous phase volume fraction (Eq. 4.49). This phase volume fraction loop is executed twice, then the phase viscosities are calculated (Eqs. 4.25, 4.26). Next the $\mathscr{A}_i$ equations (Eq. 4.35) are constructed and the temperature (Eq. 4.50) is solved. Next the PISO loop is executed twice, which consists of predicting the phase fluxes (Eq. 4.38), solving the pressure equation (Eq. 4.42), correcting the phase fluxes (Eq. 4.39), and finally reconstructing the phase velocities. Finally, the mass transfer rates are calculated (Eqs. 4.16, 4.18, and 4.51), which are under relaxed with an under relaxation factor of 0.25 to aid convergence. This entire procedure is repeated until (Eq. 4.52) is satisfied.

4.4 Drag at High Dispersed Phase Volume Fraction

Our model of boiling emulsions includes the assumptions that the emulsion is dilute and that the two dispersed phases do not impose forces on each other directly. However, even in a dilute emulsion there may be regions where the local volume fraction of one of the dispersed phases approaches one. For example, when $\rho_d > \rho_c$ the droplets will tend to settle onto horizontal surfaces and accumulate there. When such an occurrence is possible it is important that the numerical code handle it gracefully.

The most basic requirement for the numerical code at high dispersed phase fractions is stability. The code should continue to function as the volume fraction of a dispersed phase approaches one. Clearly the model described in the previous sections of this chapter fails this test. Equations (4.35) and (4.38) contain the continuous phase volume fraction in the denominator and so become undefined as $\varepsilon_c \rightarrow 0$. The offending terms arise from the drag forces on the dispersed phases. The interfacial forces, then, must be handled differently at high dispersed phase fractions so that the phase volume fraction does not appear in the denominator of any term.

A more ambitious goal is that the model should reflect the physical phenomena that occurs at large dispersed phase fractions. For example, liquid–vapor two-phase flow undergoes a series of phase transitions as the volume fraction of the vapor phase increases. As described by Wallis (1969), bubbly flow at low vapor fraction gives way to churn flow in the range $0.1 < \varepsilon_b < 0.3$, and annular flow or droplet flow is generally observed at very high vapor fractions. These transitions are strongly affected by the presence of impurities in the system. In fact, the addition of a foaming agent to a liquid–vapor system can cause bubbly flow to persist even for vapor fractions near one. The resulting foam also exhibits significant non-Newtonian behavior. Modeling such phase transitions is well beyond the scope of this study and in fact no general model exists for predicting the structure of multiphase flow under different conditions.

Rusche (2002) describes a modeling approach for two-phase flows at large dispersed phase volume fraction that models a single phase transition at a defined dispersed phase fraction. He introduces a continuous phase indicator function X_i that is defined such that $X_i \rightarrow 1$ as $\varepsilon_i \rightarrow 1$ and $X_i \rightarrow 0$ as $\varepsilon_i \rightarrow 0$. The drag force between the phases i and j of a two-phase mixture can be expressed as,

$$\mathbf{F}_{ij} = 18\mu_{\text{eff}} \mathbf{U}_{r,ij} \left(X_i \frac{\varepsilon_j}{d_j^2} - X_j \frac{\varepsilon_i}{d_i^2} \right). \tag{4.53}$$

In a two-phase mixture, $\varepsilon_j = 1 - \varepsilon_i$ and $X_j = 1 - X_i$. In the limit of $\varepsilon_i \rightarrow 1$ or $\varepsilon_j \rightarrow 1$, Eq. (4.53) becomes equal to Eq. (2.8). Use of this equation requires that a

characteristic diameter be defined even for the nominally continuous phase.
Rusche (2002) notes that the simplest definition of the continuous phase indicator
function, $X_i = \varepsilon_i$, can lead to significant errors in the terminal velocity of the
dispersed phase elements when the characteristic diameters differ greatly. Rusche
avoids this possibility by using a stronger function of the phase fraction,

$$X_i = \frac{1 + \tanh[20(\varepsilon_i - 0.5)]}{2}. \tag{4.54}$$

This approach is easily extended to mixtures of more than two phases. Equation
(4.53) can be applied directly to each pair of phases in the mixture, but defining a
continuous phase indicator function similar to Eq. (4.54) becomes more difficult. If
the relation $X_i = \varepsilon_i^2$ is adopted, the drag force between each pair of phases in
boiling emulsions is,

$$\mathbf{F}_{bc} = 18\varepsilon_b\varepsilon_c\mu_{\text{eff}}\mathbf{U}_{r,bc}\left(\frac{\varepsilon_b}{d_c^2} + \frac{\varepsilon_c}{d_b^2}\right),$$

$$\mathbf{F}_{bd} = 18\varepsilon_b\varepsilon_d\mu_{\text{eff}}\mathbf{U}_{r,bd}\left(\frac{\varepsilon_b}{d_d^2} + \frac{\varepsilon_d}{d_b^2}\right), \tag{4.55}$$

$$\mathbf{F}_{bd} = 18\varepsilon_b\varepsilon_d\mu_{\text{eff}}\mathbf{U}_{r,bd}\left(\frac{\varepsilon_b}{d_d^2} + \frac{\varepsilon_d}{d_b^2}\right).$$

These equations can be used in place of Eq. (2.8) in Eqs. (4.35) and (4.38) to
prevent the simulation from becoming unstable if $\varepsilon_c \to 0$. It is noteworthy that this
approach allows the bubble and droplet phases to exert drag forces on each other
directly, which should not occur. However, as long as the volume fraction of both
dispersed phases remains small, the force between them will be small compared to
the force exerted by either dispersed phase on the continuous phase.

References

Beard KV (1976) Terminal velocity and shape of cloud and precipitation drops aloft. J Atmos Sci
     33:851–864
Beattie DRH, Whalley PB (1982) A simple two-phase frictional pressure drop calculation
     method. Int J Multiph Flow 8:83–87
Blander M, Katz JL (1975) Bubble nucleation in liquids. AIChE J 21:833–848
Bouré JA, Delhaye JM (1982) General equations and two-phase flow modeling. In: Hetsroni G
     (ed) Handbook of multiphase systems. Hemisphere, Washington D.C
Bulanov NV, Gasanov BM (2006) Characteristic features of the boiling of emulsions having a
     low-boiling dispersed phase. J Eng Phys Thermophys 79:1130–1133
Bulanov NV, Gasanov BM (2008) Peculiarities of boiling of emulsions with a low-boiling
     disperse phase. Int J Heat Mass Transf 51:1628–1632
Bulanov NV, Skripov VP, Gasanov BM, Baidakov VG (1996) Boiling of emulsions with a low-
     boiling dispersed phase. Heat Transf Res 27:312–315
Buyevich YA (1992) Heat and mass transfer in disperse media—II. Constitutive equations. Int J
     Heat Mass Transf 35:2453–2463

Haber S, Hetsroni G (1971) The dynamics of a deformable drop suspended in an unbounded stokes flow. J Fluid Mech 49:257–277

Kays WM, Crawford ME, Weigand B (2005) Convective heat and mass transfer, 4th edn. McGraw-Hill, Boston

Maxwell JC (1904) A treatise on electricity and magnetism, 3rd edn. Clarendon, Oxford

Mori YH, Inui E, Komotori K (1978) Pool boiling heat transfer to emulsions. J Heat Transf Trans ASME 100:613–617

Ostrovskiy NY (1988) Free-convection heat transfer in the boiling of emulsions. Heat Transf Sov Res 20:147–153

Patankar SV (1980) Numerical heat transfer and fluid flow. McGraw-Hill, New York

Peskin RL (1982) Turbulent fluid-particle interaction. In: Hetsroni G (ed) Handbook of multiphase systems. Hemisphere, Washington D.C

Pinksy M, Khain A, Shapiro M (2001) Collision efficiency of drops in a wide range of Reynolds numbers: effects of pressure on spectrum evolution. J Atmos Sci 58:742–764

Pope SB (2000) Turbulent flows. Cambridge University Press, Cambridge

Roesle ML, Kulacki FA (2010) Boiling of dilute emulsions—toward a new modeling framework. Ind Eng Chem Res 49:5188–5196

Rusche H (2002) Computation fluid dynamics of dispersed two-phase flows at high phase fractions. Ph.D. Thesis, University of London, London

Silva LFLR, Lage PLC (2007) Implementation of an Eulerian multi-phase model in OpenFOAM and its application to polydisperse two-phase flows. In: Proceedings of the OpenFOAM international conference. http://www.opensourcecfd.com/conference2008/2007/index.php?option=com_content&task=view&id=3&Itemid=30. Accessed 6 Feb 2010

Taylor GI (1932) The viscosity of a fluid containing small drops of another fluid. Proc Royal Soc Lond Ser A 138:41–48

Wallis GB (1969) One-dimensional two-phase flow. McGraw-Hill, New York

Chapter 5
Measurement of Heat Transfer Coefficients in a Boiling Emulsion

Keywords Pool boiling · Heated wire · Dilute emulsion · Experimental heat transfer · Visualization · Sub-cooled boiling · Attached bubble · Superheat

Recently completed experiments investigate the behavior of dilute emulsions boiling on a small diameter heated wire. The emulsions are FC-72 in water and pentane in water. The objective is to measure the heat transfer coefficient at different temperatures for a pool of the emulsion at 1 atm with boiling on a small diameter wire. At very low droplet volume fractions, visual observation of the boiling droplets is also possible. Details of the experiments are given by Roesle and Kulacki (2012a, b) and Roesle (2010a, b).

5.1 Experimental Design and Procedure

The test cell is a chamber constructed of clear acrylic with a copper wire stretched between two bus bars located near its center (Fig. 5.1). The wire has length 100 mm and diameter 101 μm (38 AWG copper wire, L/d ≈ 1,000) and is heated by direct electrical current, which is adjusted to vary the surface temperature of the wire.

The test cell is 203 × 203 × 25.4 mm (depth), allowing it to hold ~1 L of emulsion. The depth of the cell is small enough that observation through the test cell is possible for the very dilute emulsions used in the experiments. Because the experiments are fairly short in duration (<10 min), the total energy dissipated by the wire during each experiment is small, and the bulk temperature of the emulsion remains nearly constant. The wire has nearly a constant temperature along its length, and only a minor correction is necessary to account for the temperature variation near the ends. The wire is essentially isothermal throughout its cross-section.

The voltage difference along the heated wire is measured using sense wires that contact the heated wire to avoid confounding effects of resistance in the junctions between the heated wire and supporting bus bars. The current through the wire is

M. L. Roesle and F. A. Kulacki, *Boiling Heat Transfer in Dilute Emulsions*,
SpringerBriefs in Thermal Engineering and Applied Science,
DOI: 10.1007/978-1-4614-4621-7_5, © The Author(s) 2013

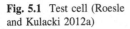

Fig. 5.1 Test cell (Roesle and Kulacki 2012a)

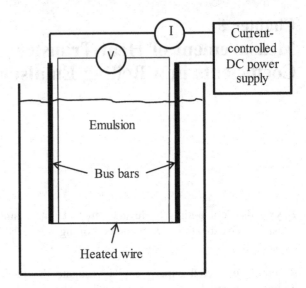

calculated from measurements of the voltage difference across a calibrated current sense resistor. At currents higher than approximately 3 A, it is possible to make a slow continuous adjustment of the current without affecting the accuracy of the measurements. The resistance of the wire, and therefore its temperature, is calculated using the voltage difference along the wire and current. These two values also determine the heat generation rate in the wire and surface heat flux. Data reduction and determination of the overall heat transfer coefficient are described in detail by Roesle (2010a, b). Overall heat transfer coefficients are determined by,

$$h = \frac{q''}{A_{\text{wire}}(T_{\text{wire}} - T_\infty)}. \tag{5.1}$$

Total uncertainty in the wire temperature, heat flux, and heat transfer coefficient are determined from the zero-level measurement uncertainties under the principle of superposition of errors (Topping 1962, Roesle 2010a, b). The uncertainty in the measured wire temperature is $\delta T_{\text{wire}} < 2\ °C$. The relative uncertainty of the heat transfer coefficient varies inversely with heat flux and is <10 % when $\Delta T > 12\ °C$, and falls to <3 % during boiling. At larger temperature differences the greatest contribution to uncertainty in the heat transfer coefficient is the uncertainty in the heat conduction to the ends of the wire. Measurements are recorded once per second. The bulk temperature of the emulsion is also measured at the beginning of each experiment.

Emulsions are prepared in batches of ~ 1.5 L. For each batch, distilled water is first degassed by boiling and then is cooled to the desired bulk temperature for the experiment. The dispersed component is then added and emulsified in the water by pumping the fluid through a turbulent jet. The emulsified component is not degassed separately, and each batch of emulsion is used immediately after preparation. Long term stability of the emulsion is not necessary and therefore no

Fig. 5.2 Distribution of droplet diameter for a sample of emulsion of 1 % FC-72 by volume

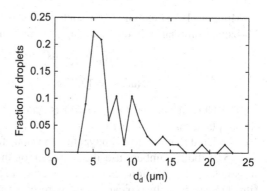

surfactants are added to the emulsion. This approach avoids any confounding effects of surfactants on nucleation and boiling.

The diameters of the droplets produced by this method are measured using photomicrography and have $4 \leq d_d \leq 22$ μm for FC-72. The average diameter is 8 μm and the volume-weighted average is 10 μm. Droplet size is not normally distributed; the greatest number of droplets have $5 \leq d_d \leq 7$ μm, and less than ten percent have $d_d > 14$ μm. The average diameter and distribution of d_d do not change significantly with volume fraction of the dispersed component, ε. The diameter is also not found to be a function of the mixing time. The distribution of measured droplet diameters for an exemplary sample of FC-72 in water is shown in Fig. 5.2.

Experiments with emulsions of FC-72 are performed at two nominal bulk temperatures, 25 and 44 °C (31 and 12 °C of sub-cooling of the FC-72, respectively). Emulsions of pentane are tested at 24 °C only (12 °C of sub-cooling of the pentane). Emulsions are prepared with $\varepsilon = 0.1$, 0.2, 0.5, and 1 %. Previous experiments have shown that for pool boiling of emulsions, the heat transfer coefficient depends on ε only up to ~ 1 % (Bulanov et al. 1996). For experiments with emulsified FC-72, the heat flux from the wire is adjusted from zero to 2 MW/m^2, and for the emulsified pentane experiments heat flux up to 7.7 MW/m^2 is used. These ranges of heat flux encompass single phase natural convection, boiling of the dispersed component only, and boiling of the dispersed and continuous components. Burnout of the wire is observed only with emulsified pentane at heat flux greater than 4 MW/m^2.

5.2   Measured Heat Transfer Coefficients

5.2.1   Water

Before performing experiments with emulsions, the heat transfer coefficient in free convection to water is measured. These experiments provide a base-line for comparison to the emulsion results and confirm that the apparatus behaves as expected.

Morgan (1975) examines 64 studies of natural convection heat transfer from horizontal circular cylinders and suggests the following correlations,

$$\begin{aligned}
\text{Nu}_{\text{film}} &= 0.675\text{Ra}^{0.058}, \quad 10^{-10} < \text{Ra} < 10^{-2}, \\
\text{Nu}_{\text{film}} &= 1.02\text{Ra}^{0.148}, \quad 10^{-2} < \text{Ra} < 10^2.
\end{aligned} \tag{5.2}$$

The Nusselt number ($\text{Nu} = hd/k$) is related to the ratio of convective to conductive heat transfer at a surface, and the Rayleigh number ($\text{Ra} = g\beta(T_s - T_\infty)d^3/\nu\alpha$) is related to the ratio of buoyant and viscous forces in buoyancy-driven flow. In Eq. (5.2), both numbers use the diameter of the heated cylinder as the characteristic length of the flow, and fluid properties in both numbers are evaluated at the film temperature. In typical heated wire experiments the Rayleigh number falls into the range $10^{-2} < \text{Ra} < 10^2$. Morgan estimates that the uncertainty of these correlations is $\sim 5\,\%$.

The boiling portion of the experimental data can be compared to the correlation developed by Rohsenow (1952),

$$\frac{c_p(T_{\text{wire}} - T_{\text{sat}})}{i_{\text{fg}}\text{Pr}^s} = C_{\text{sf}}\left(\frac{q''}{\mu_f i_{\text{fg}}}\right)^{\frac{1}{3}}\left[\frac{\sigma}{g(\rho_f - \rho_g)}\right]^{\frac{1}{6}}. \tag{5.3}$$

where for copper surfaces in water $C_{\text{sf}} = 0.013$ and $s = 1$. Although this correlation is limited to saturated boiling, the effect of sub-cooling is small for nucleate boiling (Lienhard and Lienhard 2008). For saturated boiling, Eq. (5.3) typically has $\sim 25\,\%$ error in ($T_{\text{wire}} - T_{\text{sat}}$).

Heat transfer coefficients for water are shown in Fig. 5.3 for two bulk water temperatures, 23.2 and 43.4 °C. In both cases the agreement between the experimental data for single-phase free convection and Eq. (5.2) is good. For the $T_\infty = 23.2$ °C case, the difference between Eq. (5.2) and the experimental data is less than 10 % at all temperatures, and the difference between the two generally decreases with temperature. For $T_\infty = 43.4$ °C, the data are as much as 20 % greater than given by Eq. (5.2) for $T_{\text{wire}} < 50$ °C, although the experiment and correlation are within 5 % for $T_{\text{wire}} > 90$ °C. This deviation may be due to heat loss to the walls of the test cell causing a slight decrease in T_∞ between its measurement and the experimental run (approximately 1 min). The boiling heat transfer data also agree fairly well with Eq. (5.3), taking into account the large uncertainty associated with the correlation, the large degree of sub-cooling of the water, and the unsteadiness of the wire temperature.

In both cases, superheat of ~ 40 °C is required to initiate boiling, and no significant difference in the degree of superheat required is seen between the two cases. Within four–five s of the onset of boiling the wire temperature drops by 10–15 °C. The wire temperature remains essentially constant as the heat flux at the wire surface is increased further. In neither case is transition to film boiling observed, even though the heat flux at the wire surface approaches 2 MW/m².

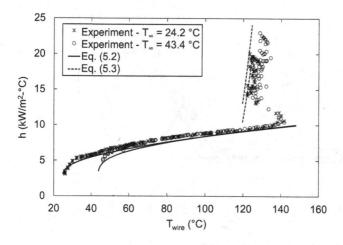

Fig. 5.3 Free convection heat transfer coefficient from heated wire to water

5.2.2 FC-72 in Water Emulsions

Heat transfer coefficients dilute emulsions of FC-72 in water are shown in Fig. 5.4. Equations (5.2) and (5.3) are included for comparison. The correlations are computed for water. Boiling curves for FC-72 alone are not included because the expected critical heat flux of FC-72 is much lower than the heat fluxes used in these experiments. Owing to the lower wire temperatures in the boiling emulsion experiments, uncertainty in the wire temperature is somewhat smaller than in the water experiments and is <1.4 °C.

Several notable trends are observed in the experimental data. First, the single-phase heat transfer coefficient is lower than for water, and the divergence from Eq. (5.2) grows larger with increasing FC-72 volume fraction and with increasing ΔT. On the other hand, boiling heat transfer is enhanced compared to water (Eq. (5.3)) and improves with increasing FC-72 volume fraction. A very large degree of superheat, relative to the saturation temperature of FC-72, is required before boiling occurs and, similar to the water experiments, a decrease in wire temperature is observed after boiling begins for experiments at low FC-72 volume fraction. Unlike the water experiments, the temperature drop does not occur immediately, but instead decreases gradually as the heat flux is increased. Although the wire temperatures become greater than the saturation temperature of water, no significant change in the heat transfer data is observed near or above 100 °C that would indicate that water suddenly begins participating in the boiling process. The wire temperature shows much less unsteadiness during boiling than in the water experiments.

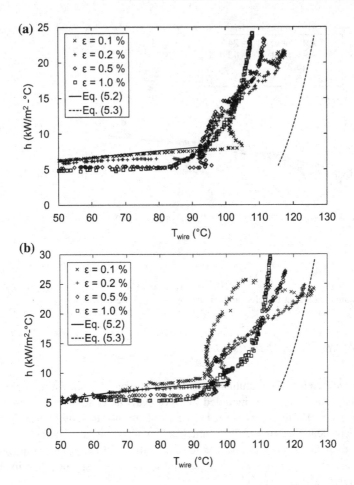

Fig. 5.4 Free convection heat transfer coefficient for heated horizontal wire in emulsions of FC-72 in water. **a** $T_\infty = 25.0 \pm 0.6$ °C, and **b** $T_\infty = 44.4 \pm 1.1$ °C. Equations (5.2) and (5.3) computed using properties of water

The experiments performed at 0.1 % FC-72 volume fraction appear to be anomalous in several ways. The degree of superheat required for boiling to begin for the low T_∞ case is the highest observed in any of the FC-72 experiments. It is also much higher than for the high T_∞ case, which is opposite the trend observed at other FC-72 volume fractions. Boiling in the high T_∞ case occurs at a lower temperature than all other experiments, but at high heat flux the surface temperature suddenly increases to near that of boiling water.

5.2.3 Pentane in Water Emulsions

In the FC-72 in water emulsions, the wire temperature must be ~ 100 °C before boiling occurs. Much of the boiling curves in Fig. 5.4 then lie above the saturation temperature of water, and it is unclear to what extent the water participates in boiling.

To address this question, a second set of experiments was performed using emulsions of pentane in water. Pentane has a saturation temperature of 35.9 °C at atmospheric pressure, approximately twenty degrees lower than FC-72. Emulsions of pentane in water should therefore boil at a lower temperature as well, so that behavior that is unambiguously due to the boiling of the pentane alone can be observed.

Figure 5.5 shows the boiling curves for pentane in water emulsions with four different volume fractions of pentane. Boiling begins at surface temperatures of ~ 75–90 °C, approximately 20 °C lower than for the FC-72 in water emulsions. Just as in the FC-72 in water emulsions, the heat transfer coefficient for single phase free convection decreases with increasing pentane volume fraction. In all cases boiling occurs at a lower temperature than in water, and at high heat flux (~ 5 MW/m^2) wire temperatures for the 0.1, 0.2, and 0.5 % pentane emulsions approach the same curve, approximately 15 °C below the temperature predicted by Eq. (5.3). This behavior is not observed for 1 % pentane because wire burnout occurs at 4 MW/m^2. For 0.5 % pentane, burnout occurs at 6.2 MW/m^2, and for 0.2 % at 7.2 MW/m^2. For 0.1 % pentane, a maximum heat flux of 7.7 MW/m^2 is achieved (at the maximum available electrical current) without reaching the critical heat flux. There is a clear trend toward decreasing critical heat flux with increasing pentane volume fraction.

For 0.2, 0.5, and 1.0 % pentane, a temperature overshoot is observed before boiling begins. For the 0.2 % case, after boiling begins the wire temperature decreases gradually as heat flux increases, which is similar to the behavior of the FC-72 in water emulsions. For the 0.5 and 1.0 % cases, the temperature overshoot is larger, and the decrease in temperature when boiling begins is abrupt, which is closer to the behavior observed in the water experiments. The temperature overshoot in the 0.5 % cases is especially large, and it appears that the wire surface temperature reaching the saturation temperature of water is the trigger that initiates boiling of the pentane.

If the portions of Fig. 5.5 in which only pentane boils are examined more closely (Fig. 5.6), some similarities to the experimental data of Bulanov et al. become apparent. The boiling heat transfer coefficient increases linearly with surface temperature or exhibits some downward curvature, similar to Figs. 2.5 and 2.7. The data here has more scatter between adjacent data points in each experiment than that of Bulanov et al. (Fig. 2.5). It may be that each data point in Fig. 2.5 represents an averaged value of a sequence of several individual measurements. In our data, there is no clear trend with pentane volume fraction of either heat transfer coefficient or the temperature at inception of boiling. Fluid combinations used in this study are different from those investigated by Bulanov et al., and other experimental conditions differ as well, so identical results are not expected.

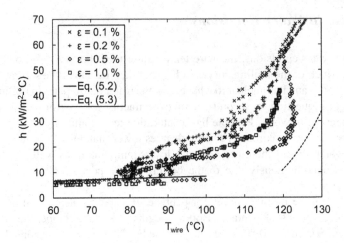

Fig. 5.5 Free convection heat transfer coefficient for heated horizontal wire in emulsion of pentane in water with $T_\infty = 24 \pm 1$ °C. Equations (5.2) and (5.3) computed using values for water (Roesle and Kulacki 2012a)

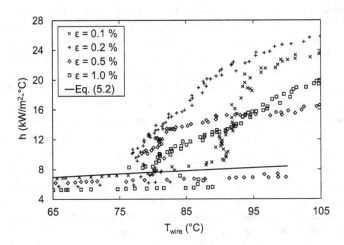

Fig. 5.6 Boiling heat transfer coefficient for heated horizontal wire in emulsion of pentane in water. Equation (5.2) is computed using properties of water

5.3 Visualization

Images of boiling described are obtained using optics with a resolution of 4.5 μm per pixel at a rate of 30 s^{-1}. A short exposure time is used to capture rapid motion of bubbles, but the rolling shutter of the camera causes image artifacts for fast-moving objects. Thus it is not possible to observe individual droplets using this imaging system due partially to the small size of the droplets ($4 < d_d \leq 22$ μm, average $d_d \approx 8$ μm), even though bubbles with $d_d \approx 20$ μm are visible. Exposure times are chosen to be as short as possible while yet allowing the image sensor to collect enough light to resolve features in the flow, and were typically between 100 and 400 μs.

An additional difficulty in observing droplets is the small difference in index of refraction between the components of the emulsions (FC-72 in water and pentane in water). Because all the fluids used in this study are transparent, droplets and bubbles are visible only due to refraction or reflection of light at their surfaces. Both FC-72 and pentane have indices of refraction close to that of water (~ 1.3). The index of refraction of their vapors is essentially unity, and thus bubbles are much more visible than droplets. Additional details of the optical system are given by Roesle (2010a, b).

5.3.1 Boiling of Water

In Fig. 5.7 several images of the heated wire are presented that were recorded during a boiling water experiment. The low contrast between bubbles and the background causes the grainy appearance of some images and lack of detail in the wire itself. The small bright circle at the center of each bubble is a reflection of the light source. The boiling curve is shown in Fig. 5.8, and the points at which the images are recorded are noted.

The heat transfer coefficient in the single phase region is slightly higher than predicted by Eq. (5.2), and the boiling heat transfer coefficient is quite close to values given by Eq. (5.3). The only notable difference is that a smaller degree of superheat is required before boiling begins. In this experiment the wire reaches 125 °C before boiling begins, rather than 140 °C as observed in the earlier experiments. The reason for this difference may be that the water used in this experiment was handled somewhat more after being degassed and before the experiment began. The water, therefore, may have reabsorbed some air before the experiment, and this may have promoted nucleation on the wire due to desorption of the air on the wire surface.

The first sign of boiling in the heat transfer data is a sudden increase in the heat transfer coefficient above the value predicted by Eq. (5.2), which corresponds to the sudden appearance of bubbles on the heated wire (Fig. 5.7b). Before this point there is no visible activity on the heated wire (Fig. 5.7a). The bubbles initially

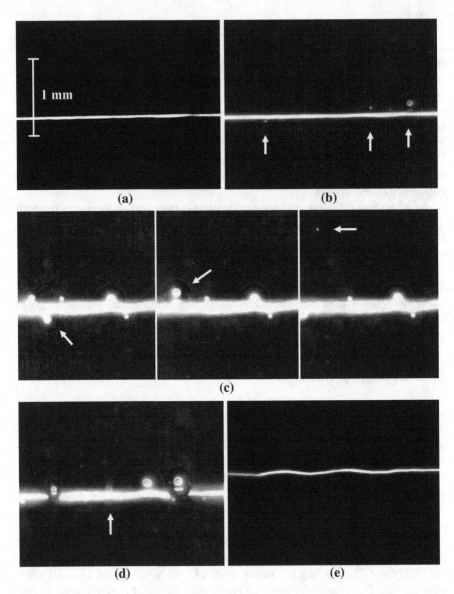

Fig. 5.7 Images of heated wire during boiling in water: **a** wire in single-phase region, **b** wire at onset of boiling, *arrows* denote bubbles attached to wire, **c** sequence of *three frames* showing bubble departure near *left edge* of frame, and **d** and **e** image artifacts caused by rapid bubble motion and vibration of wire (Roesle and Kulacki 2012b)

remain attached to the heated wire and are steady. As the heat flux dissipated by the wire increases, the bubbles grow larger and some begin to detach from the wire (Fig. 5.7c). The bubbles depart from the wire with increasing frequency with increasing heat flux.

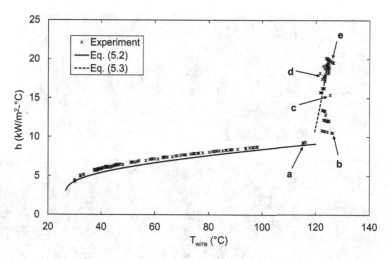

Fig. 5.8 Free convection heat transfer coefficient from heated wire to water, $T_\infty = 26.2$ °C. *Letters* denote *images* in Fig. 5.7 (Roesle and Kulacki 2012b)

At higher heat flux (>1.7 MW/m²) the vapor bubbles grow quickly enough that some artifacts become visible in the images. In Fig. 5.7d, a portion of a bubble is visible that formed while the image was being recorded. In this experiment, the camera was oriented such that the image sensor scanned rows of pixels left-to-right. Similar artifacts show in most other images. As heat flux approaches 2 MW/m², the wire begins to vibrate intermittently, probably due to the rapid growth of bubbles on the wire and their subsequent collapse after they travel into the sub-cooled liquid outside the thermal boundary layer. The vibration of the wire manifests as waviness in the recorded images (Fig. 5.7e). The wire vibrates at approximately 270 Hz with a maximum magnitude of ∼45 μm. The vibration does not have any apparent effect on the heat transfer rate from the wire.

5.3.2 Boiling of 0.1 % FC-72 in Water Emulsion

Figure 5.9 shows several images of boiling on the heated wire in an emulsion of 0.1 % FC-72 by volume in water, and Fig. 5.10 the boiling curve for the same experiment. The bulk temperature of the emulsion is 35 °C. This temperature was chosen to bisect the temperatures of the emulsions in Fig. 5.4, and the heat transfer coefficient generally does fall between the two earlier experiments. In particular the sudden increase in wire temperature at high heat flux observed in the $T_\infty = 45$ °C case also occurs here but is less significant.

The images recorded in this experiment bear many similarities to those recorded of boiling water. Bubbles first become visible on the heated wire at the same time that the heat transfer data shows the first sign of boiling (Fig. 5.9a).

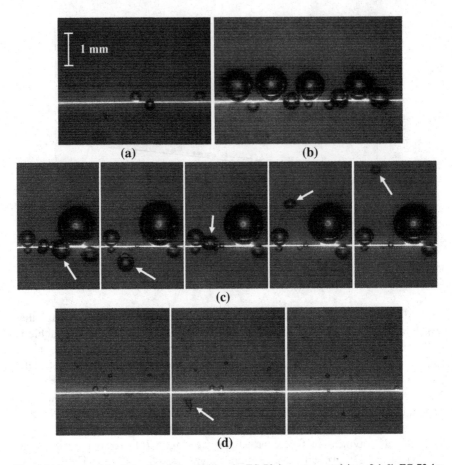

Fig. 5.9 Images of heated wire during boiling in FC-72 in water emulsion, 0.1 % FC-72 by volume: **a** onset of boiling; **b** attached bubbles at higher heat flux; **c** rapid bubble detachment; and **d** boiling at high heat flux, average bubble rise velocity is 0.0087 m/s (Roesle and Kulacki 2012b)

As heat flux increases more bubbles form, grow larger (Fig. 5.9b), and detach from the wire with increasing frequency. An interesting behavior observed in this experiment is that some bubbles depart the wire with significant velocity. Figure 5.9c shows a bubble, initially attached to the near side of the wire, departing from the wire and initially travelling downwards before rising in front of the wire due to buoyancy. The bubble also shrinks visibly owing to condensation as it rises out of the frame. This rapid departure of bubbles from the wire is likely responsible for the vibration in the wire noted in Fig. 5.7e.

At high heat flux bubbles that nucleate on the heated wire detach at a much smaller diameter than at lower heat flux (Fig. 5.9d). Due to the high temperature of the wire it is not clear whether the bubbles that nucleate on the wire are FC-72 or water. The detached bubbles visible in Fig. 5.9d have $50 < d_d < 100$ μm, and so could be the result of boiling of individual FC-72 droplets with $10 < d_d < 20$ μm.

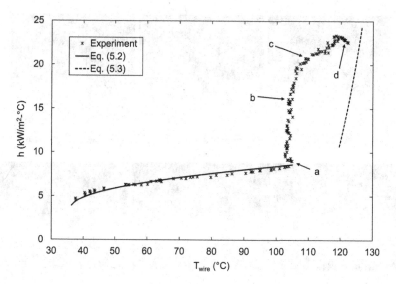

Fig. 5.10 Heat transfer coefficient for heated wire to FC-72 in water emulsion, 0.1 % FC-72 by volume, $T_\infty = 35$ °C. *Letters* denote *images* in Fig. 5.9. Correlations are calculated for water (Roesle and Kulacki 2012b)

As is seen in the second frame of Fig. 5.9d, some bubbles that nucleate on the wire surface are propelled downward, so the dispersed bubbles seen throughout the frame could be the result of this process instead. (The elongation of the bubble in the second frame is an artifact of the rolling shutter in the image sensor.) In either case, it seems likely that processes are occurring that are either too small or too fast for the camera to capture, for the small number of bubbles observed in Fig. 5.9d could not be responsible for such a large change in the heat transfer coefficient.

5.3.3 Boiling of 0.2 % FC-72 in Water Emulsion

Images of boiling of an emulsion of 0.2 % FC-72 in water are shown in Fig. 5.11. Because the opacity of the emulsions increases with the volume fraction of the dispersed component, these images could only be obtained by moving the heated wire very close to the wall of the test chamber closest to the camera. The proximity of the wall to the heated wire affects the heat transfer data and eventually the wire becomes obscured by bubbles that stick to the wall.

The appearance of bubbles coincides with the onset of boiling. The unique behavior observed here is that the bubbles do not have a steady diameter. In the previous experiments bubbles formed on the wire soon after boiling begins have a steady size that increases with heat flux. Although the trend of increasing bubble size with heat flux holds, the bubbles always fluctuate in size (Fig. 5.11a).

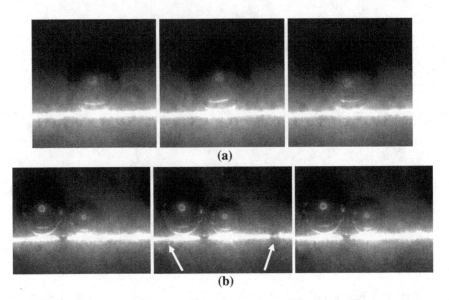

Fig. 5.11 Boiling in 0.2 % FC-72 in water emulsion. **a** Unsteady bubble at low heat flux, field of view is 1 × 1 mm. **b** Rapid boiling, field of view is 2 × 1.5 mm

At higher heat flux, the bubbles begin to detach from the wire with increasing frequency. In Fig. 5.11b two small bubbles can be observed as they pass in front of the wire but are not visible in either of the preceding or following frames. At high heat flux, vibration of the wire is observed at ∼250 Hz, similar that in boiling water.

5.3.4 Boiling of 0.1 % Pentane in Water Emulsion

Figure 5.12 shows several images of the heated wire for in boiling of 0.1 % pentane in water with 0.1 % pentane by volume. The emulsions of pentane are generally less cloudy than the emulsions of FC-72, possibly due to the closer match of index of refraction between pentane and water than between FC-72 and water. (At 20 °C, indices of refraction of water, FC-72, and pentane are 1.333, 1.251, and 1.357, respectively.) Heat transfer data are shown in Fig. 5.13.

Many of the behaviors observed in the 0.1 % pentane in water emulsion are similar to those of the 0.1 % FC-72 in water emulsion. The first bubbles become visible on the heated wire at the onset of boiling (Fig. 5.12a). As the heat flux increases, the bubbles grow larger and more bubbles form (Fig. 5.12b). Some oscillation in the size of these large bubbles is observed, but at a lower frequency than in the 0.2 % FC-72 in water experiment (Fig. 5.11a). As the heat flux increases further the bubbles begin to depart from the wire. Bubbles depart with increasing frequency and at smaller sizes as the heat flux increases (Fig. 5.12c).

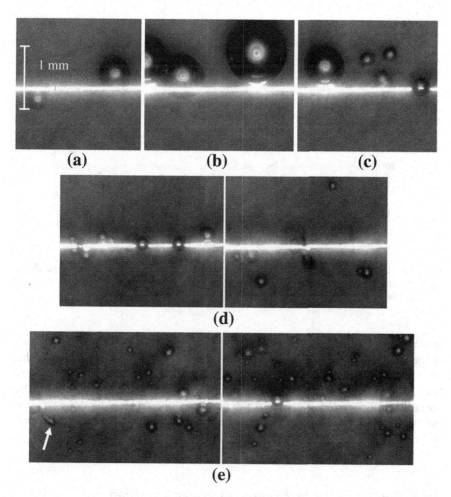

Fig. 5.12 Images of heated wire during boiling in 0.1 % pentane in water emulsion: **a** and **b** large bubbles attached to wire. **c** Departure of bubbles at higher heat flux. **d** Simultaneous departure of bubbles, average bubble rise velocity is 0.0076 m/s. **e** boiling at high heat flux, average bubble rise velocity is 0.012 m/s (Roesle and Kulacki 2012b)

At intermediate heat flux, the wire temperature fluctuates by several degrees at ∼ 108 °C with a period of a few seconds. While this temperature fluctuation occurs, many small bubbles form and depart simultaneously (Fig. 5.12d). Temperature fluctuations cease when the bubbles depart from the wire individually rather than all at once (Fig. 5.12e).

At high heat flux ($q'' > 4$ MW/m^2) bubbles with $25 < d_d < 200$ μm are observed attached to the wire, as well as in the liquid surrounding it (Fig. 5.12e). The smallest visible bubbles, if they contain only pentane, contain the same amount of pentane as a droplet with $d_d = 5$ μm and therefore could be the result of droplet boiling. At the high surface temperatures that accompany these conditions,

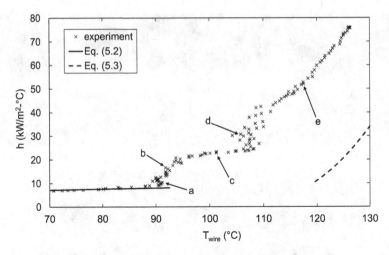

Fig. 5.13 Heat transfer coefficient for heated horizontal wire in pentane in water emulsion, 0.1 % pentane by volume, $T_\infty = 25$ °C. *Letters* denote *images* in Fig. 5.12. Equations (5.2) and (5.3) calculated using properties of water (Roesle and Kulacki 2012b)

however, the bubbles could also contain water vapor. On the other hand, the water in the emulsion is sub-cooled to such an extent ($\sim$75 °C) that it is unlikely that water vapor could exist in bubbles at any great distance from the wire. Some bubbles are observed below the heated wire Fig. 5.12e, but a bubble can also be observed in the lower left corner of the first frame moving rapidly downward, so the bubbles below the wire could all be those propelled off of the wire, was observed in the FC-72 in water emulsions (Fig. 5.9d).

5.3.5 Boiling of 0.2 % Pentane in Water Emulsion

Figures 5.14 and 5.15 show images of the heated wire and heat transfer data, respectively, for boiling in an emulsion of 0.2 % pentane in water. Large bubbles form on the wire at the inception of boiling and tend to grow and shrink erratically, similar to the 0.2 % FC-72 in water emulsion but at a slower rate. In addition, small bubbles continually form and collapse on the heated wire (Fig. 5.14a). It is not known whether the bubbles collapse entirely or become too small to observe against the heated wire. The disappearance of the bubbles cannot be caused by their departure, for the velocity of the emulsion is not high enough to transport the bubbles out of the field of view between frames.

As heat flux increases, bubbles depart from the heated wire at increasing frequency and at smaller diameters. In some cases, bubbles apparently briefly form

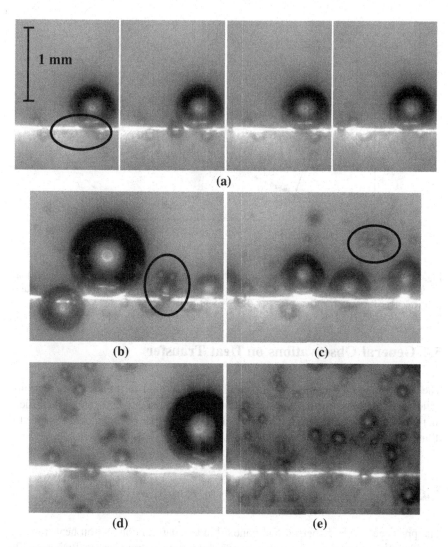

Fig. 5.14 Images of heated wire during boiling in pentane in water emulsion, 0.2 % pentane by volume. **a** onset of boiling. **b–e** boiling at increasing heat flux. Average rise velocity of bubbles is **b** 0.0052 m/s, **c** 0.0065 m/s, **d** 0.0086 m/s (Roesle and Kulacki 2012b)

small clusters without coalescing (Fig. 5.14b, c). At high heat flux the emulsion near the heated wire appears similar to that for the 0.1 % pentane emulsion (Fig. 5.14e), although the number density of bubbles is higher, which is to be expected when the bubbles form from pentane droplets. The range of bubble sizes in Fig. 5.14e is essentially the same as in Fig. 5.12e.

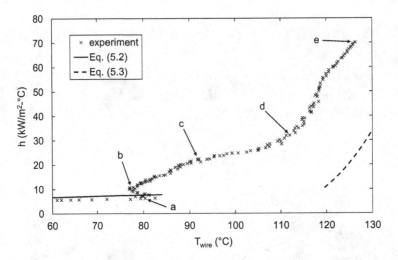

Fig. 5.15 Heat transfer coefficient for heated horizontal wire in pentane in water emulsion, 0.2 % pentane by volume, $T_\infty = 23.5$ °C. Letters denote images in Fig. 5.14. Equations (5.2) and (5.3) calculated using properties of water (Roesle and Kulacki 2012b)

5.4 General Observations on Heat Transfer

The wide range of behaviors of the emulsions described in the previous sections defies easy, detailed explanation. Nevertheless trends in the data can be identified that provide some insight into the physical processes that occur in heat transfer to dilute emulsions.

5.4.1 Single Phase Heat Transfer in Emulsions

The presence of the dispersed component hinders natural convection heat transfer significantly. There is a trend towards lower heat transfer coefficient with increasing disperse phase volume fraction, although there is some scatter in the data for pentane in water emulsions (Fig. 5.16). Some decrease in the heat transfer rate is expected because pentane and FC-72 both have much lower thermal conductivity than that of water ($k = 0.595$, 0.117, and 0.056 W/m °C for water, pentane and FC-72 respectively at 25 °C). According to the effective medium theory of Maxwell (1904), at a dispersed phase volume fraction of 1 % the effective conductivity of the emulsion should be no more than 1.5 % lower, even if the dispersed phase is a perfect insulator. The effect of the dispersed component on mixture viscosity mixture is similarly small (Taylor 1932). The large decrease in the single-phase heat transfer coefficient is strong evidence that droplets of the dispersed component collect on the surface of the heated wire. If the dispersed

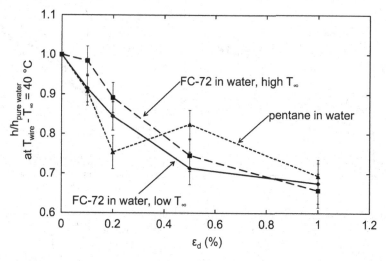

Fig. 5.16 Decrease in single-phase free convection heat transfer coefficient for emulsions

liquid forms a uniform layer on the wire surface, the layer need only be a few microns thick to cause the observed decrease in heat transfer. It is not possible to determine with the visual techniques employed here how much of the surface of the wire is covered and whether the droplets coalesce into a continuous layer of liquid on the wire.

5.4.2 Boiling of the Continuous Component Versus Dispersed Component Boiling

Emulsions of FC-72 in water begin boiling close to the saturation temperature of water, so that it is difficult to separate the effects of boiling FC-72 droplets alone from evaporation of the water. In contrast, the boiling curves for emulsions of pentane in water have two distinct regions. At surface temperatures below $\sim 105\ ^\circ$C only the pentane is boiling, and the boiling curves show similar behavior to that reported by Bulanov et al. (Fig. 5.6). At higher surface temperatures, it is likely that the water boils as well, which is enhanced by the presence of pentane bubbles. For the 0.1, 0.2, and 0.5 % volume fraction cases there is a distinct jump in surface temperature between these two regions.

Having noted the two distinct boiling regimes for pentane in water emulsions, it is reasonable to return to the FC-72 emulsion data and look for similarities. Figure 5.17 compares boiling curves for the pentane and FC-72 emulsions as a function of the degree of superheat of the dispersed component. The figure contains the warm FC-72 data, as the degree of sub-cooling of the dispersed component is similar to that of the pentane emulsion experiments. When compared

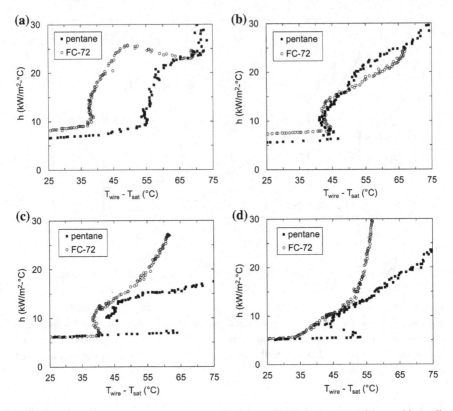

Fig. 5.17 Comparison of pentane in water emulsions to FC-72 in water emulsions with small sub-cooling. Emulsions are **a** 0.1 %, **b** 0.2 %, **c** 0.5 %, and **d** 1.0 % dispersed component by volume (Roesle and Kulacki 2012a)

on this basis, the boiling curves for 0.2, 0.5, and 1.0 % dispersed component volume fraction emulsions for $40 < T_{wire} - T_{sat} < 50$ °C range are remarkably similar. The only exception is the 0.1 % case, in which the behavior of the FC-72 emulsion has already been noted as being anomalous. The boiling curves are expected to diverge for $T_{wire} - T_{sat} > 50$ °C because, for the FC-72 in water emulsions, this temperature range corresponds to surface temperatures high enough to cause the water to begin boiling as well. This result suggests that the most significant effects of the bubbles of the dispersed component are caused by their presence, and the properties of the vapor in the bubbles are unimportant.

5.4.3 Surface Temperature Overshoot

A striking difference between pentane and FC-72 emulsions in Fig. 5.17c, d is the large temperature overshoot that occurs before the pentane emulsions begin boiling.

In fact, such overshoots occur inconsistently throughout the experiments. No correlation with degree of sub-cooling of the emulsion or the dispersed component volume fraction is apparent. One clear distinction is that the largest temperature overshoots are seen in the pentane in water emulsion data, and those cases the temperature of the wire drops suddenly after the inception of boiling rather than gradually as heat flux increases. The inconsistent data suggests that the temperature overshoots are linked to some aspect of the preparation of the emulsions that was not adequately controlled.

A likely source of variability in the emulsions is in the degree to which they are degassed. The water used in the emulsions is first degassed by boiling, but the dispersed component is not separately degassed. Additionally, after degassing the water is cooled to room temperature and handled further in the course of producing the emulsion. This procedure provides opportunity for the water to re-absorb atmospheric gasses. The large degree of superheat required to initiate boiling in the water experiments (Fig. 5.3) suggests that the emulsions remain at least partially degassed, but differences in the handling of each batch of emulsion can result in different amounts of dissolved gasses in each experiment. It is noteworthy that none of the previous investigators of boiling in emulsions mention degassing procedures in their experiments, and Bulanov's theory of chain boiling depends on the presence of dissolved atmospheric gases in the dispersed component. The data here suggests that dissolved gases play a role in the initiation of boiling but do not impact the boiling heat transfer coefficient.

5.4.4 Attached Bubbles

The models reported in Chaps. 2 and 3 rely on boiling of droplets of the dispersed component in the thermal boundary layer around a heated surface and not on the surface itself.

The droplets of the emulsions used in these experiments have $4 < d_d < 22$ µm and, if they boil individually, would produce bubbles with $20 < d_d < 130$ µm. The images obtained of boiling emulsions at high heat flux (Figs. 5.14b–e, 5.12e, 5.9d) capture bubbles with diameters as small as ~ 25 µm. While individual droplets cannot be seen, most bubbles that result from boiling droplets are visible. It is clear from the images that at low heat flux there are no small bubbles in the emulsion around the heated wire.

Instead of dispersed bubbles, at low heat flux large bubbles form on the wire and remain attached to it. The first appearance of the bubbles in the recorded video coincides with the inception of boiling in the heat transfer data, and it is reasonable to conclude that the former cause the latter. These large bubbles merit further consideration. The bubbles are observed to grow from the wire, so they probably formed at nucleation sites on the wire that were wetted by the droplets that had collected on the wire. Further growth of the bubbles might be enabled by coalescence of droplets that flow past the wire with the bubbles. This result directly contradicts the statement by Bulanov et al. (2009) that boiling of the dispersed component at nucleation sites on the heated surface occurs at surface temperatures above the saturation temperature of the dispersed component but causes no significant change in heat transfer coefficient until dispersed boiling begins to occur. Despite that, it will be shown in a following section that the boiling curves obtained for the experiments described here are qualitatively and quantitatively similar to boiling curves reported by Bulanov et al. (2006) for oil in water emulsions.

The bubbles contact only a small fraction of the wire's total surface area, yet they cause a large rise in the heat transfer coefficient. The presence of the bubbles certainly changes the flow field around the wire, but because the bubbles are mostly stationary and located above the wire (while the emulsion rises past the wire from below), it is not likely that the bubbles themselves would cause significant disruption of the thermal boundary layer around the wire. Phase change and circulation inside the bubbles might account for the improved heat transfer, however. Each bubble contains vapor of the dispersed component, which condenses at the top of the bubble surface where the temperature of the emulsion is less than the saturation temperature of the dispersed component. A film of the dispersed component liquid therefore grows on the surface of the bubble, and this film flows down the sides of the bubble. When the liquid reaches the vicinity of the wire it evaporates again, and the vapor circulates back to the top of the bubble. Thus the bubbles might function as small heat pipes to increase the rate of thermal energy transport out of the thermal boundary layer surrounding the wire. This explanation, it must be noted, is speculation at this point.

Clearly any model based on boiling of individual droplets around the heated surface is at best incomplete. It appears however, that for the pentane emulsions the small dispersed bubbles start forming at a lower surface temperature in the 0.2 % case than in the 0.1 % case. If that trend continues at higher dispersed component fractions (where video cannot be obtained), then boiling of dispersed droplets would be an important boiling mechanism for higher fraction emulsions. For the 0.2 % pentane emulsion, many small dispersed bubbles are observed at wire temperatures below 100 °C, so they must be pentane vapor. It is tempting to attribute the higher boiling heat transfer coefficient of the 0.2 % emulsion as compared to the 0.1 % emulsion for $T_{wire} < 100$ °C (Fig. 5.6) to the presence of these dispersed bubbles, which should improve heat transfer as described in Chap. 4. However, this explanation does not account for the fact that the heat transfer coefficients of the 0.5 and 1.0 % pentane emulsions are lower in the same temperature range.

5.4.5 Pressure Jump Due to Interfacial Tension

It is important to note the effect of interfacial tension on the saturation temperature of the droplets. Because of interfacial tension between the droplets and the continuous component, the pressure in the droplets is higher than ambient pressure by $2\sigma_{dc}/r_d$.

This pressure jump causes the saturation temperature of the droplets to be higher than expected based upon the ambient pressure.

In order to evaluate the change in saturation temperature, the interfacial tension between the droplets and continuous component must be known. At 22 °C, the interfacial tension between pentane and water is 0.051 N/m, which falls between the surface tension of water and pentane (0.072 and 0.0158 N/m, respectively) (Goebel and Lunkenheimer 1997). Similar data is not available for FC-72 and water, but because the surface tension of FC-72 is also much lower than that of water (0.0115 N/m at 20 °C), it is reasonable to assume that the interfacial tension between FC-72 and water is also lower than the surface tension of water. The surface tension of water may be used to estimate an upper bound on the increase of saturation temperature of the droplets.

The FC-72 droplets in this study have $4 \leq d_d \leq 22$ μm. Using the surface tension of water at 80 °C (0.063 N/m), the pressure jump in the droplets is between 11 and 63 kPa. In both FC-72 and pentane emulsions this pressure rise causes an increase in saturation temperature of 3–15 °C. Thus all of the droplets in the emulsions are superheated well before any boiling is observed.

5.4.6 Comparison to Earlier Experiments

The results described above can be compared to those of Bulanov et al. (2006). They measure boiling heat transfer coefficients for emulsions of several combinations of fluids on heated horizontal and vertical wires. The most comparable experiments to the present study are for emulsions of diethyl ether in water and R-113 in water. Earlier studies have shown that for dilute emulsions, the behavior of water in oil and oil in water emulsions differ, so only the results by Bulanov et al. for water-based emulsions are considered here. In the following discussion, the experiments are grouped by the degree of sub-cooling of the emulsion compared to the saturation temperature of the emulsified component.

Due to the range of saturation temperatures of the dispersed components of the emulsions under consideration, the boiling curves are presented as a function of the degree of superheat of the heated wire, $T_{\text{wire}} - T_{\text{sat}}$. In addition, the Bulanov et al. study used wires of varying diameters and orientations, so the single-phase heat transfer coefficient differs considerably between experiments. To compare the effects of the boiling emulsion in different experiments, the increase in the heat transfer coefficient above its single-phase value, $h - h_{1\text{ph}}$, is used. Some of the results reported by Bulanov et al. do not include single-phase convection, so the value of $h_{1\text{ph}}$ is taken as the first (lowest-temperature and lowest heat transfer coefficient) data point for these experiments. Bulanov et al. do not make any mention of degassing their emulsions.

Figure 5.18 compares heat transfer coefficients for emulsions with a small degree of sub-cooling of approximately 12 °C. These include the pentane in water and the warm FC-72 in water emulsions described above, as well as three diethyl

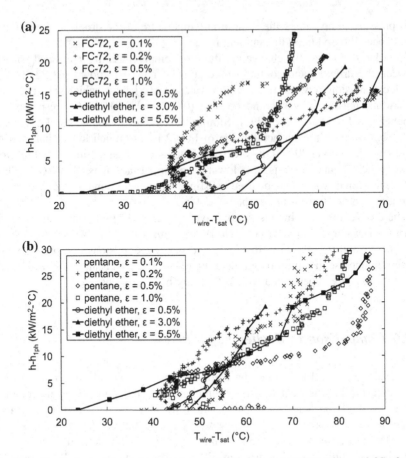

Fig. 5.18 Comparison of boiling heat transfer in water-based emulsions with $\sim 12\,°C$ of sub-cooling between diethyl ether (Bulanov et al. 2006) and **a** FC-72, and **b** pentane

ether in water emulsions tested by Bulanov et al. (2006). The diethyl ether in water emulsions contained no surfactants and had an average droplet diameter of 60 μm. The $\varepsilon = 0.5$ and 3.0 % emulsions were tested with a vertical platinum wire with 100 μm diameter and 50 mm length. The 5.5 % diethyl ether emulsion was tested with a horizontal platinum wire with 50 μm diameter and 36 mm length. The difference in wire orientation and diameter appears to correspond to a difference in the slope of the boiling curve and the degree of superheat required to initiate boiling.

The behavior of the 5.5 % diethyl ether emulsion is similar to that of the 1.0 % FC-72 emulsion for wire superheat up to 50 °C (Fig. 5.18a). At that degree of superheat, the water in the FC-72 in water emulsions is superheated as well and may begin to participate in the boiling. However, diethyl ether has a lower saturation temperature of 34.6 °C at 1 atm pressure, close to that of pentane, and so the water is not expected to participate in boiling until $T_{wire} - T_{sat}$ reaches ~ 70 °C.

As seen in Fig. 5.18b, emulsions whose emulsified components have similar saturation temperatures, also have similar boiling heat transfer coefficients. Except for low degrees of superheat, the boiling heat transfer coefficients of the diethyl ether emulsion and the pentane emulsions with the highest dispersed component volume fraction (5.5 and 1.0 % respectively) are quite close to each other. The boiling heat transfer coefficients of the more dilute diethyl ether emulsions (0.5 and 3.0 %) mostly fall between those of the more dilute pentane emulsions (0.1 and 0.2 %), and the curves have approximately the same slope for superheats up to 60 °C. The more dilute emulsions do show opposite trends with volume fraction, however: the diethyl ether emulsion with $\varepsilon = 0.5$ % has a higher boiling heat transfer coefficient than the 3.0 % emulsion, while the 0.1 % pentane emulsion has a lower boiling heat transfer coefficient than the 0.2 % emulsion.

Figure 5.19 compares the boiling heat transfer coefficients of emulsions with a higher degree of sub-cooling of ~ 25 °C. The results of the cool FC-72 in water emulsions described above are included along with those of emulsions of R-113 in water (Bulanov et al. 2006). The experiments with R-113 emulsions were performed using a vertical heated platinum wire with 50 μm diameter and 46 mm length, and the emulsions had an average droplet diameter of 60 μm.

Despite these differences in heated surface diameter and orientation, as in the previous comparison the emulsions with high volume fractions (0.5 and 1.0 % for pentane and 1.0 and 2.0 % for R-113) have very similar boiling heat transfer coefficients. The reason for the large drop in wire temperature for the 0.5 % R-113 experiment is unknown and is not found in any other experiments. R-113 has a saturation temperature of 48 °C at 1 atm which is close to that of FC-72.

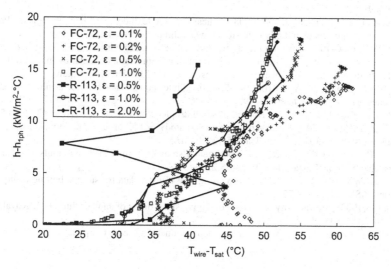

Fig. 5.19 Comparison of boiling heat transfer of FC-72 in water emulsions and R-113 in water emulsions (Bulanov et al. 2006) with ~ 24 °C sub-cooling (Roesle and Kulacki 2012a)

What is most interesting in these results is that despite the differences in heated wire diameter, length, orientation, and material, and the differences in properties of the dispersed components, droplet sizes, and volume fractions, many of the emulsions show very similar behavior. Although the data base is still much too sparse to make any general conclusions or correlations, it appears that for emulsions of low-boiling-point liquids in water, the most important property of the dispersed component is its saturation temperature. Although not shown in detail here, water-in-oil emulsions display a much more gradual increase in boiling heat transfer coefficient with temperature (Fig. 2.5, Bulanov and Gasanov 2008). This result suggests that the properties of the continuous component are important to boiling heat transfer. Or, there may be relatively weak dependence on the relative properties of the dispersed and continuous components. For all the oil-in-water emulsions examined here, $k_c > k_d$ and $c_{p,c} > c_{p,d}$, while the opposite is true for water-in-oil emulsions.

References

Bulanov NV, Gasanov BM (2008) Peculiarities of boiling of emulsions with a low-boiling disperse phase. Int J Heat Mass Transf 51:1628–1632

Bulanov NV, Gasanov BM, Turchaninova EA (2006) Results of experimental investigation of heat transfer with emulsions with low-boiling disperse phase. High Temp 44:267–282

Bulanov NV, Gasanov BM, Muratov GN (2009) Critical volume and chain activation of boiling sites in emulsions with low-boiling dispersed phase. High Temp 47:864–869

Bulanov NV, Skripov VP, Gasanov BM, Baidakov VG (1996) Boiling of emulsions with a low-boiling dispersed phase. Heat Trans Res 27:312–315

Goebel A, Lunkenheimer K (1997) Interfacial tension of the water/n-alkane interface. Langmuir 13:369–372

Lienhard JH IV, Lienhard JH V (2008) A heat transfer textbook, 3rd edn. Phlogiston Press, Cambridge http://web.mit.edu/lienhard/www/ahtt.html. Accessed 13 Nov 2009

Maxwell JC (1904) A treatise on electricity and magnetism, 3rd edn. Clarendon, Oxford

Morgan VT (1975) The overall convective heat transfer from smooth circular cylinders. In Irvine TF, Hartnett JP (ed) Advances in heat transfer vol 11. Elsevier, Amsterdam

Roesle ML, Kulacki FA (2010a) Boiling of dilute emulsions—toward a new modeling framework. Ind Eng Chem Res 49:5188–5196

Roesle ML, Kulacki FA (2010b) Boiling of small droplets. Int J Heat Mass Trans 53:5587–5595

Roesle ML, Kulacki FA (2012a) An experimental study of boiling in dilute emulsions, part A: heat transfer. Int J Heat Mass Trans 55:2160–2165

Roesle ML, Kulacki FA (2012b) An experimental study of boiling in dilute emulsions, part B: visualization. Int J Heat Mass Trans 55:2166–2172

Rohsenow WM (1952) A method of correlating heat transfer data for surface boiling of liquids. Trans ASME 74:969–975

Taylor GI (1932) The viscosity of a fluid containing small drops of another fluid. Proc Royal Soc Lond Ser A 138:41–48

Topping J (1962) Errors of observation and their treatment, 3rd edn. Chapman and Hall, London

Chapter 6
Simulation of Boiling in a Dilute Emulsion

Keywords Eulerian multiphase model · Finite volume · CFD · Collision efficiency · Pseudo-turbulence · Thermal boundary layer · Free convection · Boiling
heat transfer · Dilute emulsion

Two-dimensional simulations of a boiling emulsion using the algorithm of Chap. 4
are discussed in this chapter. The symmetrical computational domain is illustrated
in Fig. 6.1a. A symmetry boundary condition is imposed at the vertical axis that
passes through the centerline of the wire, and a no-slip boundary condition is
imposed at the wire surface and at the outer edge domain. The wire temperature is
held constant and the temperature at the outer edge of the domain is fixed at T_∞.
The initial temperature field is uniform at T_∞, and the initial velocity for each
component is uniform at zero. The initial bubble volume fraction is zero and initial
droplet volume fraction is set equal to the average dispersed component volume
fraction for the emulsion. The domain is split into a mesh of hexahedral cells with
125 cells in the radial direction and 32 cells in the angular direction. Cells are
clustered along the radius with a factor of 167, i.e., cells adjacent to the outer
boundary of the domain have a radial length 167 times larger than cells adjacent to
the wire. No clustering is imposed in the angular direction (Fig. 6.1b). The mesh is
generated using the standard meshing utility of OpenFOAM^{TM}.

Simulations were performed on different meshes to ensure that grid indepen-
dence is achieved. Reducing the number of cells by 25 % in each direction (to 96
by 24 cells) results in changes in the heat transfer coefficient by ~ 1 %, while
increasing the number of cells by 25 % changes the heat transfer coefficient by less
than 0.2 %. These tests were carried out under conditions that produce the
strongest temperature and volume fraction gradients in the emulsion.

The duration of each simulation is $4 < t < 10$ s. Simulations with a small
temperature difference between T_{wire} and T_∞ require longer durations to achieve
steady state. In all simulations, a plume of fluid rises from the heated wire and
eventually reaches the top of the domain. Once the plume reaches the top, the size
of the domain begins to have an effect on the heat transfer from the wire, and the
results are no longer indicative of the behavior of the emulsion around the wire

M. L. Roesle and F. A. Kulacki, *Boiling Heat Transfer in Dilute Emulsions*,
SpringerBriefs in Thermal Engineering and Applied Science,
DOI: 10.1007/978-1-4614-4621-7_6, © The Author(s) 2013

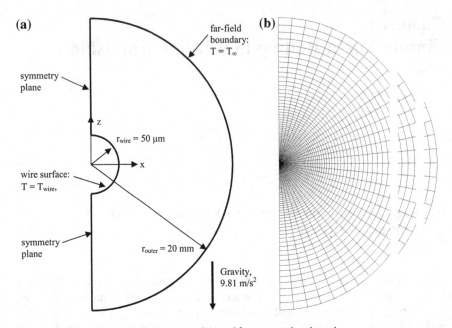

Fig. 6.1 **a** Simulation domain (not to scale), and **b** computational mesh

alone. For large temperature differences this occurs after ~ 4 s (Fig. 6.2a). For all simulations, this duration is sufficient for the heat transfer rate from the wire to reach a steady value.

Simulations are presented here for water and emulsions of FC-72 in water. Average FC-72 volume fractions are 0.1, 0.2, 0.5, 1.0, and 2.0 %. Bulk temperatures are 28 and 43 °C, and $48 < T_{wire} < 108$ °C. Fluid properties are evaluated at the film temperature. Parameter variation, such as collision efficiency, pseudo-turbulent viscosity, and thermal conductivity are investigated as well.

6.1 Water

Typical results for the temperature field and velocity fields for water are shown in Figs. 6.2 and 6.3. Results are for the end of the simulation at $t = 4$ s. At this time the plume of heated water has just reached the top of the domain, but temperature and velocity fields in the vicinity of the wire are nearly unchanged for $t > 1$ s. Due to the small wire diameter and low fluid velocity near the wire, no detached vortices or vortex shedding from the wire are expected, nor is such behavior observed. The assumption that the flow field is symmetrical is therefore reasonable. Two stagnation points exist at the expected locations, the top and bottom of the wire.

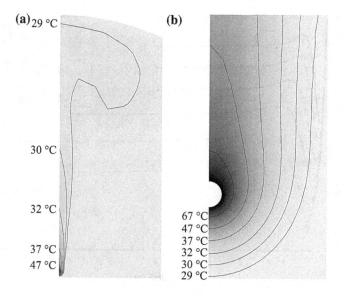

Fig. 6.2 Temperature contours for a horizontal wire in water, $T_{wire} = 98\ °C$, $T_\infty = 28\ °C$, $\varepsilon_d = 0$. **a** top half of domain. **b** vicinity of the wire

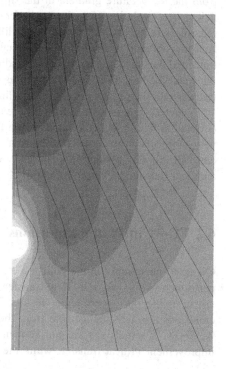

Fig. 6.3 Streamlines for water. $T_{wire} = 98\ °C$, $T_\infty = 28\ °C$, $\varepsilon_d = 0$. Shading indicates velocity magnitude. The maximum velocity is 0.0077 m/s at upper left corner

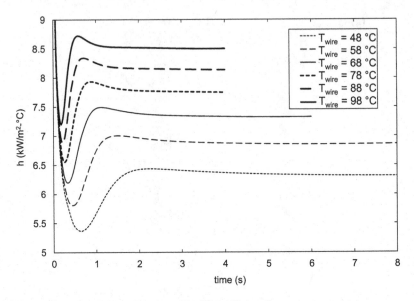

Fig. 6.4 Predicted evolution of the heat transfer coefficient for water

As each simulation progresses, the heat flux from the wire surface is calculated from the temperature gradient at the surface to determine the heat transfer coefficient. The evolution of the heat transfer coefficient over time takes the same form regardless of the overall temperature difference (Fig. 6.4). The very high initial heat transfer rate is a result of the initial condition, $T = T_\infty$ everywhere in the fluid, and thus the large temperature gradient at the surface. As energy diffuses into the water the heat transfer rate decreases until buoyant flow commences. The heat transfer rate then increases again, and finally settles to its steady value.

Figure 6.5 compares results to the experimental data and the Morgan (1975) correlation with good agreement within the scatter of the experimental data. This very close match is actually somewhat surprising, as it has been found that the constant property assumption can lead to significant errors in simulations of flow and heat transfer in water (Kumar et al. 2007).

6.2 FC-72 in Water Emulsions

Owing to uncertainty in the collision efficiency (Sect. 4.2), initial simulations are performed with $\eta = 0.03$. Heat transfer coefficients are shown in Fig. 6.6 and do not resemble the experimental results. The results for the 0.1, 0.2, and 0.5 % emulsions are indistinguishable from those of water. The results for the 0.1 and 0.2 % FC-72 emulsions are not shown. The 1 % emulsion shows only a slight improvement in heat transfer, while the 2 % emulsion has a lower heat transfer

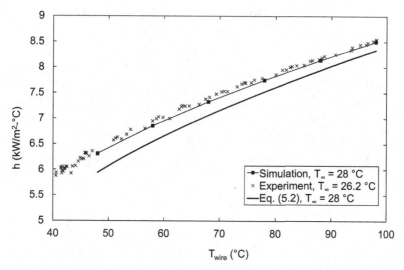

Fig. 6.5 Heat transfer coefficients for heated horizontal wire in water

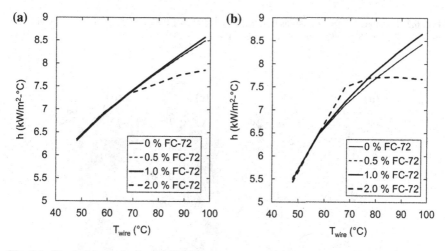

Fig. 6.6 Simulation results for FC-72 in water. **a** $T_\infty = 28\,°C$, **b** $T_\infty = 43\,°C$

coefficient than water. In all cases the effects of the emulsified FC-72 are more pronounced for smaller subcooling of the bulk temperature. The reasons for this behavior can be discerned by examining the temperature and volume fraction fields around the heated wire.

Shown in Figs. 6.7 and 6.8 are the volume fractions of bubbles and droplets respectively in the region around the heated wire. Figure 6.9 shows the mass transfer rate near the heated wire. The solid line in the figures is the 56.3 °C isotherm, the saturation temperature of FC-72. In all cases except for the 2 % FC-72 emulsion, not much boiling of the dispersed component occurs. The FC-72

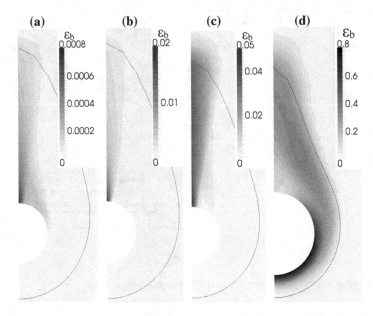

Fig. 6.7 Simulated ε_b for emulsions of **a** 0.1 %, **b** 0.5 %, **c** 1.0 %, and **d** 2.0 % FC-72 by volume in water. $T_\infty = 28$ °C. $T_{\text{wire}} = 98$ °C. Line is the $T = 56.3$ °C isotherm

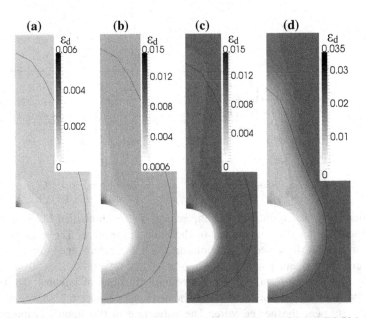

Fig. 6.8 Simulated ε_d for emulsions of **a** 0.1 %, **b** 0.5 %, **c** 1.0 %, and **d** 2.0 % FC-72 in water. $T_\infty = 28$ °C, $T_{\text{wire}} = 98$ °C, line is the $T = 56.3$ °C isotherm

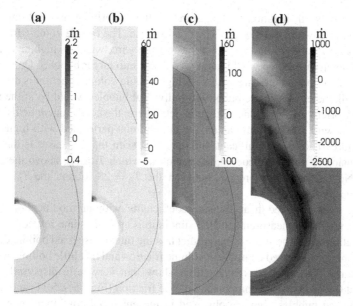

Fig. 6.9 Simulated $\dot{m}$, in kg/m^3 s, for emulsions of **a** 0.1 %, **b** 0.5 %, **c** 1.0 %, and **d** 2.0 % FC-72 in water. $T_\infty = 28$ °C, $T_{wire} = 98$ °C, and line is the $T = 56.3$ °C isotherm

droplets tend to settle onto the top half of the wire surface, where some of them boil. Some droplets accumulate at the stagnation point at the top of the wire, and the rate of boiling due to wall contact is highest there. For emulsions with less than 0.5 % FC-72, the area immediately above the wire is the only location where any boiling is observed. For emulsions with 0.5 % or more FC-72 by volume, boiling due to collisions between bubbles and droplets occurs in the plume above the wire as the bubbles rise through the boundary layer. However, the rate of boiling due to collisions is a very strong function of the droplet volume fraction and a strong function of the bubble volume fraction (Eqs. 4.13 and 4.16). Boiling that occurs in the plume above the heated wire is far from the wire and has little effect on the overall heat transfer from the wire.

Only for emulsions of 2 % FC-72 is rapid boiling seen all around the wire surface (Figs. 6.7d, 6.8d, 6.9d). In this case, the droplet volume fraction is high enough to sustain boiling around the bottom half of the wire even though no boiling by wall contact occurs in this region. Boiling occurs throughout the region where $T > T_{sat,FC-72}$, and the most rapid boiling occurs near the outer edge of that region, rather than at the wire surface as at lower FC-72 volume fractions. The volume fraction of FC-72 droplets within the thermal boundary layer is depleted significantly. However, as Fig. 6.6 shows, this rapid boiling does not improve the heat transfer coefficient. Instead, the high volume fraction of bubbles around the wire reduces the thermal conductivity of the emulsion (Eq. 4.27), effectively insulating the wire in a manner similar to film boiling and overwhelms the increase in thermal conductivity due to agitation of the emulsion predicted by (Eq. 4.24).

It is also worth mentioning that some of the underlying assumptions of the model break down in the 2 % FC-72 emulsion case. The bubble volume fraction is as high as 0.75 around the bottom surface of the wire, which is high enough that the bubbles in this region would rapidly coalesce into larger bubbles. On the other hand, the layer below the wire with very high bubble volume fraction is no thicker than the diameter of a single bubble, so individual bubbles would be in the region only very briefly. Although not shown in detail here, the ε_b, ε_d, and $\dot{m}$ fields around the heated wire are very similar for the simulations performed with higher bulk temperature. The most significant difference between the two cases is that when the emulsion is not as subcooled, the region in which $T > T_{sat}$ above the wire is larger. No qualitative differences between the $T_b = 28$ °C and the $T_b = 43$ °C cases were observed.

While the predicted heat transfer coefficients without modification do not compare well with measurements, the simulations agree in some respects with the visualization data. The simulations predict that the rate of dispersed boiling depends strongly on the dispersed component volume fraction and that little boiling occurs at low volume fraction. The visualizations show that very little dispersed boiling occurs at 0.1 % dispersed component volume fraction (Figs. 5.9, 5.11) and most of the dispersed bubbles seen are observed in the enhanced continuous component boiling regime. For the 0.2 % pentane emulsion, many more dispersed bubbles are observed, and at lower temperatures than for the 0.1 % emulsions. Thus, the simulations and the visualizations agree regarding the dependence of the rate of dispersed droplet boiling on dispersed component volume fraction.

6.3 Pseudo-Turbulent Effects

It has long been argued that the heat transfer enhancement in boiling emulsions is due in part to agitation of the continuous component by rapid boiling of the dispersed droplets. In Chap. 4, a model is suggested for quantifying the effects of this agitation. It is based on the displacement of liquid surrounding a boiling droplet due to expansion and considers only displacement due to droplet growth to the maximum size achieved by the vapor bubble during boiling. However, after a highly superheated droplet boils, the bubble oscillates (Chap. 3). Although this result neglects acoustic damping, thermal damping which is included is more significant (Plesset and Prosperetti 1977). The liquid around a boiling droplet therefore undergoes many repeated displacements inwards and outwards. It follows that the agitation of the emulsion is more significant than the model predicts.

Effects of bubble oscillation on the pseudo turbulence in the model of boiling emulsions can be explored by modifying Eqs. (4.23) and (4.24) to include an additional factor,

$$\mu_{c,T} = 0.1\, K_T \dot{m} \frac{\rho_c}{\rho_b} R_b{}^2, \tag{6.1}$$

$$k_T = 0.1\, K_T \dot{m} c_{p,c} \frac{\rho_c}{\rho_b} R_b{}^2. \tag{6.2}$$

Simulated boiling curves for $K_T = 1$, 10, 100, 300, and 1,000 are shown in Figs. 6.10 and 6.11. Results for emulsions of 0.1, 0.2, and 0.5 % FC-72 are not shown because heat transfer coefficients are indistinguishable from those of water. Results for 2.0 % FC-72 and $K_T = 1000$ are not shown because of difficulty in achieving computational stability.

These results show that an improvement in heat transfer coefficient can be achieved if the agitation of the emulsion is more effective than predicted by the model in Chap. 4. The behavior of the emulsion is a very strong function of the volume fraction of the dispersed component, however, which does not agree with the experimental results. The simulations show an improvement in heat transfer coefficient starting at a much lower temperature than in the experiments, a consequence of assuming that droplets will boil upon contact with any surface where $T > T_{\text{sat}}$. Use of a correlation to predict the nucleation temperature of the droplet liquid on the heated surface would bring that aspect of the simulation results into agreement with the experimental results. In fact, simulation would then predict the sudden jump in heat transfer coefficient observed in some experiments (e.g., the $\varepsilon_d = 0.5$ % curves in Fig. 5.4).

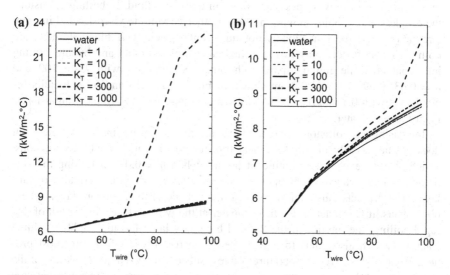

Fig. 6.10 Simulation results for heated wire in 1.0 % FC-72 in water emulsions. **a** $T_\infty = 28$ °C, **b** $T_\infty = 43$ °C

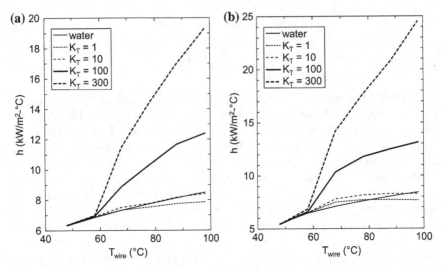

Fig. 6.11 Simulation results for heated wire in 2.0 % FC-72 in water emulsions. **a** $T_\infty = 28$ °C.
b $T_\infty = 43$ °C

6.4 Collision Efficiency

Estimates of collision efficiency discussed in Sect. 4.2 are drawn from a study of
two solid spheres moving past each other in quiescent fluid. In boiling emulsions
the droplets and bubbles moving past each other have finite viscosity and internal
circulation, and the emulsion is not quiescent especially if many droplets are
boiling. These factors may increase the chances of a droplet and bubble coming
into contact while moving past each other. Simulations with $K_T = 100$ and
$\eta = 0.03$, 0.09, 0.3, and 0.9 are shown in Figs. 6.12, 6.13 and 6.14. Results for
emulsions with 0.1 and 0.2 % FC-72 are not shown, as they are indistinguishable
from those of water.

Increasing the collision efficiency increases the rate of boiling as expected, but
does not necessarily improve the heat transfer coefficient. For 2.0 % emulsions,
which already experience significant boiling when $\eta = 0.03$, increasing the collision efficiency causes droplets to boil sooner after entering the thermal boundary
layer. The droplets thus tend to boil near the outside of the boundary layer, where
boiling has little impact on the heat transfer at the wire surface. As a result of this
rapid boiling, the interior of the thermal boundary layer has high bubble volume
fraction, which also tends to impede heat transfer. These effects are most pronounced at high surface temperature and low subcooling, when the thickness of the
region around the wire where $T > T_{sat}$ is greater. Thus, Fig. 6.14 shows that as η
increases the heat transfer coefficient increases slightly at low surface temperature
while it decreases significantly at high surface temperature.

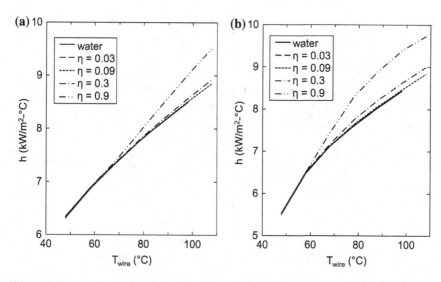

Fig. 6.12 Simulation results for heated wire in FC-72 in water emulsions. $K_T = 100$. $\varepsilon_d = 0.5$ %. **a** $T_\infty = 28$ °C. **b** $T_\infty = 43$ °C

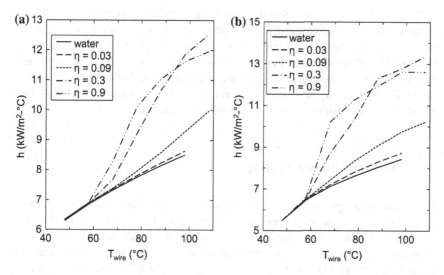

Fig. 6.13 Simulation results for heated wire in FC-72 in water emulsions. $K_T = 100$. $\varepsilon_d = 1.0$ %. **a** $T_\infty = 28$ °C. **b** $T_\infty = 43$ °C

For emulsions of 0.5 and 1 % FC-72, increasing the collision efficiency does provide an improvement in the heat transfer coefficient. As in the higher volume fraction case, the effects are more pronounced at higher surface temperature and for lower subcooling. At the highest collision efficiency, the heat transfer coefficient for the 1.0 % FC-72 emulsion begins to decrease in the same manner as for the 2.0 % FC-72 emulsion.

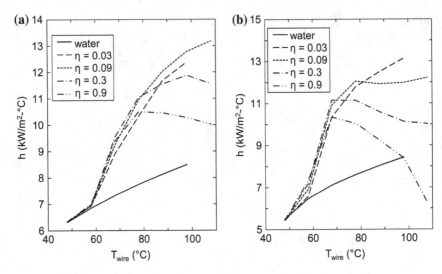

Fig. 6.14 Simulation results for heated wire in FC-72 in water emulsions. $K_T = 100$. $\varepsilon_d = 2.0$ %. **a** $T_\infty = 28$ °C, **b** $T_\infty = 43$ °C

The parameters K_T and η therefore have rather different effects on the behavior of the simulations. Increasing K_T always increases the heat transfer coefficient, but the heat transfer coefficient remains a very strong function of the dispersed component volume fraction. For a fixed value of K_T, increasing η does not significantly increase the maximum heat transfer coefficient that can be achieved by an emulsion. Instead, heat transfer is enhanced over a wider range of surface temperatures and dispersed component volume fractions.

6.5 Interfacial Force Scaling

Forces acting on droplets in an emulsion include drag, virtual mass, lift, and rotational forces. Thus far, it has been assumed that only drag forces are significant in boiling. To evaluate the validity of these assumptions, computations have been performed in which the lift, drag, virtual mass, and rotational forces were calculated using Eqs. (2.9)–(2.11) with $C_{vm} = 0.5$ and $C_L = C_R = 0.25$. The simulation conditions were $T_\infty = 28$ °C, $T_{wire} = 98$ °C, $\varepsilon_d = 2.0$ %, $\eta = 0.03$, and $K_T = 100$.

Results show that the assumptions made in Chap. 4 are justified (Fig. 6.15). The lift, rotational, and virtual mass forces are always smaller than drag forces. Of the three neglected forces, only the virtual mass force for the bubbles is within an order of magnitude of the drag force, and that occurs only near a region of rapid condensation of the bubbles above the thermal boundary layer. None of the neglected forces are significant within the boundary layer. Thus, they would have little effect on heat transfer from the heated wire.

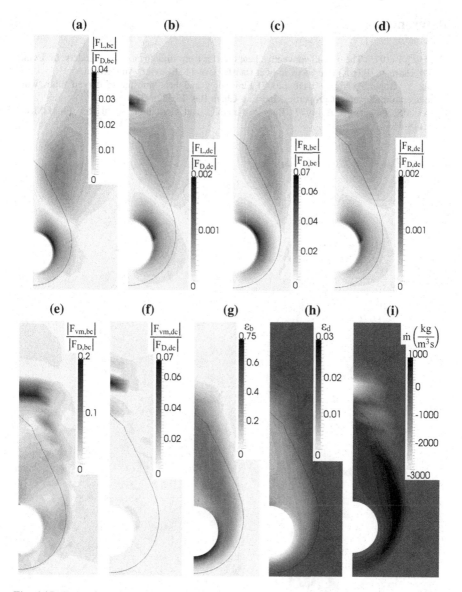

Fig. 6.15 Forces near the heated wire, $T_{\text{wire}} = 98\ °C$. $T_\infty = 28\ °C$. **a, b** Lift to drag force ratio. **c, d** rotational to drag force ratio. **e, f** virtual mass to drag force ratio. **g** ε_b, **h** ε_d, **i** $\dot{m}$. *Solid line* is the $T = 56.3\ °C$ isotherm

References

Morgan VT (1975) The overall convective heat transfer from smooth circular cylinders. In: Irvine
    TF, Hartnett JP (eds) Advances in heat transfer, vol 11. Elsevier, Amsterdam
Kumar V, Gupta P, Nigam KDP (2007) Fluid flow and heat transfer in curved tubes with
    temperature-dependent properties. Ind Eng Chem Res 46:3226–3236
Plesset MS, Prosperetti A (1977) Bubble dynamics and cavitation. Annu Rev Fluid Mech
    9:148–185

Chapter 7
Closure

Keywords Pool boiling · Boiling heat transfer · Dilute emulsion · Convection · Multiphase flow · Heat transfer modeling

Enhancement of heat transfer in dilute emulsions by boiling of the low boiling point dispersed component has been demonstrated experimentally by Bulanov and co-workers, but the subject has not otherwise received much attention. Experimental data in the literature was essentially limited to that reported by Bulanov and co-workers. Thus experiments of boiling on small diameter, heat wire have been carried out with emulsions of pentane in water and FC-72 in water, two fluid combinations that have not previously been studied. These fluid combinations have properties closer to emulsions that would have practical use in high heat flux electronics cooling applications than the water in oil and oil in oil emulsions that have been the primary subjects of previous studies.

Unlike previous studies that consider only boiling of the dispersed component of the emulsion, the experiments discussed here are extended to surface temperatures higher than the saturation temperature of the continuous component as well. Two distinct boiling heat transfer regimes are observed. At low temperatures boiling of the dispersed component is observed, while at high temperatures there is enhanced boiling of the continuous component. The transition between the two regimes occurs at ∼5–15 °C above the saturation temperature of the continuous component. During boiling of the dispersed component, the heat transfer coefficient for the emulsion increases roughly linearly with surface temperature, sometimes after an initial overshoot in surface temperature. This behavior is similar to previously reported results. When compared on the basis of superheat of the dispersed component, the data for FC-72 in water emulsions and pentane in water emulsions are very similar in the dispersed component boiling region. This result suggests that aside from saturation temperature, the presence of the dispersed component is more important than its material properties. During enhanced boiling of the continuous component, the heat transfer coefficient increases much more rapidly with temperature and either parallels or slowly converges toward the

M. L. Roesle and F. A. Kulacki, *Boiling Heat Transfer in Dilute Emulsions*,
SpringerBriefs in Thermal Engineering and Applied Science,
DOI: 10.1007/978-1-4614-4621-7_7, © The Author(s) 2013

behavior predicted by the Rohsenow correlation (Eq. 5.3), $q'' \propto (T - T_{sat})^3$. However, the 0.1 % FC-72 emulsion is anomalous in this regard.

The most remarkable result of the visualization experiments is the presence of large attached bubbles on the heated wire. The formation of the bubbles coincides with the inception of boiling as seen in the heat transfer data. Also remarkable is the absence of any visual evidence of individual droplet boiling at very low dispersed component fractions and low temperatures. The large attached bubbles represent an additional boiling mode that has not been reported by previous workers and, when few individual droplets boil, it represents the dominant mode of boiling heat transfer.

Earlier studies of boiling emulsions have mostly been limited to reporting experimental results and trends in the data. The only previous attempt at formulating a model of the mechanisms at play in boiling dilute emulsions was put forward by Bulanov and co-workers. The Bulanov model contains several assumptions, some of which are incompatible with each other. Consequently, in the current work a new model has been developed.

A one-dimensional model of a single boiling droplet in superheated liquid is used to simulate the behavior of droplets under the conditions found in the thermal boundary layer of a boiling dilute emulsion. The simulations reveal that the droplet boils very quickly so that the resulting bubble overshoots its equilibrium diameter and oscillates. The droplet does not boil rapidly enough to create a shockwave. The data and insight gained from this simulation has been used in the larger model of boiling emulsions.

A model of boiling dilute emulsions has been developed based upon the Euler–Euler model of multiphase flows. The general balance equations developed by Drew and Passman (1999) are applied to the present situation, thus providing a rigorous and physically consistent framework. The model contains three phases: the continuous component, liquid droplets of the dispersed component, and bubbles resulting from boiling of individual droplets. Mass, momentum, and energy transfer between the phases is based upon the behavior of, and interaction between, individual elements of the dispersed components. Boiling mechanisms are modeled based upon the behavior of individual droplets and bubbles, rather than processes at the nanometer scale that are fundamentally unobservable. Droplet boiling is assumed to occur when it contacts a heated surface with $T > T_{sat}$ or a vapor bubble in a region where $T > T_{sat}$. Collision efficiency between droplets and bubbles and chain-boiling of closely spaced droplets are considered.

The model is meant to describe the behavior of emulsions in the dispersed component boiling regime and thus it does not account for phase change of the continuous component. The model also does not include the large attached bubbles that were observed in experiments. However, these effects can be added and represent possible avenues for further analysis. The model is extensible in many directions and represents a solid platform for exploration of other geometries and conditions. Simulations and parameter studies performed with a numerical implementation of the model have identified ranges of conditions in which the model predicts significant heat transfer enhancement by emulsion boiling, and several aspects of the numerical results agree qualitatively with experimental observations. The simulations do not show any heat transfer enhancement for very dilute emulsions ($\varepsilon_d \leq 0.2$ %), which correspond to conditions in which the large attached bubbles are the dominant visible mode of boiling observed in experiments. The following are several areas of research that address unresolved challenges revealed in this volume.

The large attached bubbles of the dispersed component are a new discovery and are not included in the present model. The bubbles appear to be the dominant boiling heat transfer mechanism under some circumstances, and therefore deserve closer study. It is hypothesized that the bubbles enhance heat transfer primarily by circulation within each bubble. That is, vapor condenses in cool surroundings at the top of the bubble and then flows down the side of the bubble in a thin liquid film. The liquid film evaporates at the base of the bubbles (the heated surface), and the vapor circulates back to the top. Analytical work is required to quantify this effect. Direct observation of the flow field around the bubbles, perhaps by particle image velocimetry, would also be useful to search for other effects of the bubbles.

One obstacle to accounting for the large attached bubbles in the current multiphase model is the basic assumption in the current model that the multiphase mixture is dispersed. In contrast, each large bubble is a single contiguous region of the vapor phase and is large compared to other features in the flow including the heated surface itself. To properly model the bubble, it is necessary to identify the bubble's surface and incorporate surface tension in the model, as surface tension is the dominant force in determining the bubble's shape and is responsible for the bubble remaining attached to the heated surface. If the hypothesis, stated above, that the attached bubbles enhance heat transfer essentially by acting as small heat pipes, then it is also necessary to model a thin film of liquid of the dispersed phase at the surface of each bubble and evaporation and condensation at its surface. These processes are independent of the boiling and condensation of individual droplets and bubbles in the dispersed phase, and a complete model of boiling heat transfer in dilute emulsions must incorporate both.

The present model takes a rather simplistic view of the agitation of the emulsion by the boiling of droplets, as evidenced by the need for the empirical parameter K_T. One interesting consequence of this view is that the agitation is predicted to be a strong function of the droplet size (Eqs. 4.23 and 4.24). In contrast, previous studies that have investigated droplet size have not noted a strong link between the droplet size and heat transfer. An extension of the work in Chap. 3 to simulate boiling droplets in shear flow or in the presence of other boiling droplets would be beneficial.

The current model does not include a turbulence model, and therefore it is only practical for use in laminar flows. (Direct numerical simulation of turbulence is still possible with the present model, but simulations would probably require more computing resources than is practical.) Work has been done by Rusche (2002) and others on integrating the k-epsilon turbulence model into Euler–Euler models of multiphase flow. Such a turbulence model, if adapted to the present model of boiling emulsions, would also have a benefit for the previous suggestion. It may be possible to directly link the kinetic energy possessed by a boiling droplet to the rate of generation of turbulence kinetic energy.

The current model uses an estimate for collision efficiency based upon simulations of collisions between two solid spheres. However, in boiling emulsions, droplets, bubbles, and the continuous component all have different viscosities. An analysis or series of simulations of collisions between droplets and bubbles would be beneficial. Additionally, the effects of shear or agitation in the continuous liquid on the collision efficiency should be explored.

Another aspect of collisions that deserves further attention is the outcome of collisions. In the present model, it is assumed that a collision between a droplet and bubble results in the droplet boiling. It is likely, however, that in such a case the droplet would coalesce with the bubble and result in a single larger bubble. There will also be droplet–droplet and bubble–bubble collisions that can result in coalescence. Accounting for these behaviors would require tracking size distributions for both droplet and bubble phases, which was deemed an unnecessary complication in the present model. These behaviors could also be accounted for by abandoning the Euler–Euler model and developing an Euler–Lagrange model of the emulsion. Such a model would only be usable in very small domains due to the large number density of droplets, even in very dilute emulsions. Such a model may be well suited to situations such as microchannels, however.

Additions to the current model could be made to address the effects of surfactants. Surfactants would impact the outcomes of collisions discussed in the previous suggestion. Surfactants would also change the behavior of the dispersed component at the heated surface. With the discovery of the large attached bubble boiling mechanism, additional experimental work with emulsions with surfactants, especially visual observations of the attached bubbles, would be beneficial.

Previous experimental studies have found some effects of suspended particles in the dispersed component, which were explained in the context of nucleation within the dispersed droplets. These effects should be explainable within the framework of the current model. The behavior of the particles in the large attached bubbles requires investigation as well.

Finally, our experiments, as well as most previous experimental studies, focus on heat transfer from thin heated wires. The advantage of this configuration is in the simplicity of the experimental apparatus. However, practical applications of boiling emulsions would involve free convection from flat surfaces or forced convection in small diameter channels. Further study of these configurations is required, both experimental and analytical. It should be confirmed that the behaviors observed for boiling on thin wires also occur in other configurations.

References

Drew DA, Passman SL (1999) Theory of multicomponent fluids. Springer, New York
Rusche H (2002) Computation fluid dynamics of dispersed two-phase flows at high phase fractions. Ph.D. Thesis, University of London, London

Acknowledgments

The following have graciously allowed their material to be reproduced in this monograph.

Table 2.1: Reprinted from *International Journal of Heat and Mass Transfer*, vol. 49 issue 23–24, Chen T., Klausner J. F., Garimella S. V., Chung J. N., "Subcooled boiling incipience on a highly smooth microheater," pp. 4399–4406, copyright (2006), with permission from Elsevier.

Table 3.1: Reprinted from *International Journal of Heat and Mass Transfer*, vol. 53, Roesle M. L., Kulacki F. A., "Boiling of small droplets," pp. 5587–5595, copyright (2010) with permission from Elsevier.

Figure 2.3: Reprinted from *Transactions of the ASME—Journal of Heat Transfer*, vol. 100 issue 4, Mori Y. H., Inui E., Komotori K., "Pool boiling heat transfer to emulsions," pp. 613–617, copyright (1978), with permission from ASME.

Figure 2.4: Reprinted with permission of Begell House, Inc., from *Heat Transfer—Soviet Research*, vol. 20 issue 1, Ostrovskiy N. Yu, "Free-convection heat transfer in the boiling of emulsions," copyright (1988); permission conveyed through Copyright Clearance Center, Inc.

Figures 2.5, 2.7, 2.9: Reprinted from *International Journal of Heat and Mass Transfer*, vol. 51, Bulanov N. V., Gasanov B. M., "Peculiarities of boiling of emulsions with a low-boiling disperse phase," pp. 1628–1632, copyright (2008), with permission from Elsevier.

Figure 2.6: Reprinted from *Journal of Engineering Physics and Thermophysics*, vol. 79 issue 6, Bulanov N. V., Gasanov B. M., "Characteristic features of the boiling of emulsions having a low-boiling dispersed phase," pp. 1130–1133, copyright (2006), with kind permission from Springer Science and Business Media.

Figure 2.8: Reprinted with permission from *Industrial and Engineering Chemistry Research*, vol. 49, Roesle M. L., Kulacki F. A., "Boiling of dilute emulsions—toward a new modeling framework," pp. 5188–5196, copyright (2010) American Chemical Society.

M. L. Roesle and F. A. Kulacki, *Boiling Heat Transfer in Dilute Emulsions*, SpringerBriefs in Thermal Engineering and Applied Science, DOI: 10.1007/978-1-4614-4621-7, © The Author(s) 2013

Figures 3.1, 3.9, 3.10: Reprinted from *International Journal of Heat and Mass Transfer*, vol. 53, Roesle M. L., Kulacki F. A., "Boiling of small droplets," pp. 5587–5595, copyright (2010) with permission from Elsevier.

Figures 5.1, 5.5, 5.17, 5.19: Reprinted from *International Journal of Heat and Mass Transfer*, vol. 55, Roesle M. L., Kulacki F. A., "An experimental study of boiling in dilute emulsions, part A: Heat transfer," pp. 2160–2165, copyright (2012) with permission from Elsevier.

Figures 5.7–5.10, 5.32–5.34: Reprinted from *International Journal of Heat and Mass Transfer*, vol. 55, Roesle M. L., Kulacki F. A., "An experimental study of boiling in dilute emulsions, Part B: Visualization," pp. 2166–2172, copyright (2012) with permission from Elsevier.

Appendix

Averaged Balance Equations

Multiple approaches have been used historically to derive the averaged balance equations for multiphase flows from the local equations of motion (Bouré and Delhaye 1982; Hill 1998; Wallis 1969). In this section the conditional averaging approach as used by Drew and Passman (1999) is described. Key features of this approach are the use of ensemble averaging to achieve the averaged equations in a single averaging operation, and the use of a phase indicator function to isolate each phase in the mixture.

Exact Balance Equations

The averaging process begins with the local, instantaneous balance equations for mass, momentum, and energy. These equations can be expressed as a generic balance equation

$$\frac{\partial \rho \Psi}{\partial t} + \nabla \cdot \rho \Psi \mathbf{U} = \nabla \cdot \gamma + \rho \zeta, \tag{A1}$$

where values for Ψ, γ, and ζ are given in Table A.1 (Drew and Passman 1999). For multiphase flows, these balance equations, plus appropriate jump conditions at interfaces between the phases and constitutive relations for each fluid, form a complete description of the flow. These equations are typically not used directly in numerical simulations because the computing resources required to resolve all the interfaces are prohibitive.

M. L. Roesle and F. A. Kulacki, *Boiling Heat Transfer in Dilute Emulsions*,
SpringerBriefs in Thermal Engineering and Applied Science,
DOI: 10.1007/978-1-4614-4621-7, © The Author(s) 2013

Table A.1 Variables in the generic conservation equation (Eq. A1)

Conservation principle	Ψ	γ	ζ
Mass	1	0	0
Momentum	**U**	**T**	**b**
Energy	$e + \frac{1}{2}|\mathbf{U}|^2$	$\mathbf{T} \cdot \mathbf{U} - \mathbf{q}$	$\mathbf{b} \cdot \mathbf{U} + S$

Phase Indicator Function

A basic tool used to develop the conditional averaged balance equations is the phase indicator function X_i, defined as

$$X_i(\mathbf{x}, t) = \begin{cases} 1 & \text{if phase } i \text{ is present at} (\mathbf{x}, t) \\ 0 & \text{otherwise} \end{cases} \tag{A2}$$

This function has the effect of picking out one phase of a multiphase mixture (Drew and Passman 1999). The phase indicator function is used to condition Eq. (A1) so that balance equations may be obtained for each phase individually.

The phase indicator function is thus constant except at the interface, so the gradient of X_i may be used to pick out the interfaces of phase i. Drew and Passman express ∇X_i in terms of the Dirac delta function $\delta(x, t)$,

$$\nabla X_i = \mathbf{n}_i \delta(\mathbf{x} - \mathbf{x}_\mathrm{I}, t), \tag{A3}$$

where $\mathbf{n}_i$ is the unit normal vector pointing toward phase i and $\mathbf{x}_i$ is the location of the interface. Another important property of the phase indicator function is the topological equation

$$\frac{\partial X_i}{\partial t} + \mathbf{U}_I \cdot \nabla X_i = 0, \tag{A4}$$

which can be interpreted physically to mean simply that the interface travels with the velocity of the interface $\mathbf{U}_I$ (Hill 1998).

Conditional Averaging

To obtain averaged balance equations for each phase of a multiphase mixture, the exact balance equation (Eq. A1) is first multiplied by the phase indicator function. The first three terms of the resulting equation can be expanded to obtain

$$\frac{\partial X_i \rho \Psi}{\partial t} + \nabla \cdot X_i \rho \Psi \mathbf{U} = \nabla \cdot X_i \gamma + X_i \rho \zeta + \rho \Psi \left(\frac{\partial X_i}{\partial t} + \mathbf{U} \cdot \nabla X_i \right) - \gamma \cdot \nabla X_i. \tag{A5}$$

Next, the velocity on the right-hand side of Eq. (A5) is expanded to $\mathbf{U}_I + (\mathbf{U} - \mathbf{U}_I)$ and the topological equation is applied

$$\frac{\partial X_i \rho \Psi}{\partial t} + \nabla \cdot X_i \rho \Psi \mathbf{U} = \nabla \cdot X_i \gamma + X_i \rho \zeta + [\rho \Psi (\mathbf{U} - \mathbf{U}_I) - \gamma] \cdot \nabla X_i. \qquad (A6)$$

The last term on the right-hand side represents the transfer of the quantity Ψ to phase i across its interfaces by mass transfer and by diffusive flux. The conditioned balance equation can then be averaged using ensemble averaging

$$\frac{\partial \overline{X_i \rho}}{\partial t} + \nabla \cdot \overline{X_i \rho \mathbf{U}} = \overline{\rho (\mathbf{U} - \mathbf{U}_I) \cdot \nabla X_i}, \qquad (A7)$$

$$\frac{\partial \overline{X_i \rho \mathbf{U}}}{\partial t} + \nabla \cdot \overline{X_i \rho \mathbf{U} \mathbf{U}} = \nabla \cdot \overline{X_i \mathbf{T}} + \overline{X_i \rho \mathbf{b}} + \overline{[\rho \mathbf{U}(\mathbf{U} - \mathbf{U}_I) - \mathbf{T}] \cdot \nabla X_i}, \qquad (A8)$$

$$\frac{\partial \overline{X_i \rho \left(e + \frac{1}{2}|\mathbf{U}|^2 \right)}}{\partial t} + \nabla \cdot \overline{X_i \rho \left(e + \frac{1}{2}|\mathbf{U}|^2 \right) \mathbf{U}} = \nabla \cdot \overline{X_i (\mathbf{T} \cdot \mathbf{U} - \mathbf{q})} + \overline{X_i \rho (\mathbf{b} \cdot \mathbf{U} - S)} +$$
$$\overline{\left[\rho \left(e + \frac{1}{2}|\mathbf{U}|^2 \right)(\mathbf{U} - \mathbf{U}_I) - (\mathbf{T} \cdot \mathbf{U} - \mathbf{q}) \right] \cdot \nabla X_i}, \qquad (A9)$$

An averaged balance equation for the internal energy may be obtained from Eqs. (A8) and (A9).

$$\frac{\partial \overline{X_i \rho e}}{\partial t} + \nabla \cdot \overline{X_i \rho e \mathbf{U}} = -\nabla \cdot \overline{X_i \mathbf{q}} + \overline{X_i \mathbf{T} : \nabla \mathbf{U}} + \overline{X_i \rho S} + \overline{[\rho e (\mathbf{U} - \mathbf{U}_I) + \mathbf{q}] \cdot \nabla X_i}$$
$$(A10)$$

A number of manipulations are now required to cast these equations into the more familiar forms given in Sect. 2.2 (Eqs. 2.3–2.5). First, averaged equations are defined as follows. The phase fraction is simply the average of the phase indicator function

$$\varepsilon_i = \overline{X_i} \qquad (A11)$$

As noted by Drew and Passman (1999), ε_i is often called the volume fraction of phase i, which implies volume averaging. By the definition given here, however, the volume fraction is a result of ensemble averaging and is defined even as the volume approaches zero. As noted in Sect. 2.2, density and the diffusive and fluctuation fluxes are averaged using component averages

$$\bar{\rho}_i = \frac{\overline{X_i \rho}}{\varepsilon_i}, \qquad (A12)$$

$$\overline{\mathbf{T}}_i = \frac{\overline{X_i \mathbf{T}}}{\varepsilon_i} \qquad (A13)$$

$$\bar{\mathbf{q}}_i = \frac{\overline{X_i \mathbf{q}}}{\varepsilon_i}, \qquad (A14)$$

$$\mathbf{T}_i^{\mathrm{Re}} = -\frac{\overline{X_i \rho \mathbf{U}_i' \mathbf{U}_i'}}{\varepsilon_i} \qquad (A15)$$

$$\mathbf{q}_i^{\mathrm{Re}} = \frac{\overline{X_i \rho \mathbf{U}_i' e_i'}}{\varepsilon_i} + \frac{\overline{X_i \rho \mathbf{U}_i' \frac{1}{2} |\mathbf{U}_i'|^2}}{\varepsilon_i} - \frac{\overline{X_i \mathbf{T} \cdot \mathbf{U}_i'}}{\varepsilon_i} \tag{A16}$$

The fluctuation heat flux contains fluctuation internal energy flux, fluctuation kinetic energy flux, and fluctuation shear working terms. Other quantities are Favré (mass-weighted) averaged, including velocity, internal energy, and body sources $\mathbf{b}_i$ and $\mathbf{S}_i$,

$$\overline{\mathbf{U}}_i = \frac{\overline{X_i \rho \mathbf{U}}}{\varepsilon_i \overline{\rho}_i}, \tag{A17}$$

$$\overline{e}_i = \frac{\overline{X_i \rho e}}{\varepsilon_i \overline{\rho}_i}, \tag{A18}$$

$$\overline{\mathbf{b}}_i = \frac{\overline{X_i \rho \mathbf{b}}}{\varepsilon_i \overline{\rho}_i}, \tag{A19}$$

$$\overline{\mathbf{S}}_i = \frac{\overline{X_i \rho \mathbf{S}}}{\varepsilon_i \overline{\rho}_i}. \tag{A20}$$

Equations (A7), (A8), and (A10) can be can be put in terms of these averaged variables by expressing each exact variable as the sum of its average and a fluctuating component, and then applying the Reynolds averaging rules. For example, Drew and Passman (1999) expand the momentum flux term in Eq. (A8),

$$\overline{X_i \rho \mathbf{UU}} = \overline{X_i \rho \left(\overline{\mathbf{U}}_i + \mathbf{U}_i'\right)\left(\overline{\mathbf{U}}_i + \mathbf{U}_i'\right)} = \left(\overline{X_i \rho}\right)\overline{\mathbf{U}}_i \overline{\mathbf{U}}_i + \overline{X_i \rho \mathbf{U}_i' \mathbf{U}_i'} = \varepsilon_i \overline{\rho}_i \overline{\mathbf{U}}_i \overline{\mathbf{U}}_i - \varepsilon_i \mathbf{T}_i^{\mathrm{Re}}. \tag{A21}$$

The choices of variable weightings made by Drew and Passman simplify this process considerably. In the limit of incompressible flows the distinction between component weighting and mass weighting vanishes. Other terms are handled similarly, and the results are given as Eqs. (2.3)–(2.5),

$$\frac{\partial \varepsilon_i \overline{\rho}_i}{\partial t} + \nabla \cdot \left(\varepsilon_i \overline{\rho}_i \overline{\mathbf{U}}_i\right) = \Gamma_i \tag{2.3}$$

$$\frac{\partial \varepsilon_i \overline{\rho}_i \overline{\mathbf{U}}_i}{\partial t} + \nabla \cdot \left(\varepsilon_i \overline{\rho}_i \overline{\mathbf{U}}_i \overline{\mathbf{U}}_i\right) = \nabla \cdot \left[\varepsilon_i \left(\overline{\mathbf{T}}_i + \mathbf{T}_i^{\mathrm{Re}}\right)\right] + \varepsilon_i \overline{\rho}_i \overline{\mathbf{b}}_i + \mathbf{F}_i + \mathbf{U}_{i,I}\Gamma_i \tag{2.4}$$

$$\frac{\partial \varepsilon_i \overline{\rho}_i \overline{e}_i}{\partial t} + \nabla \cdot \left(\varepsilon_i \overline{\rho}_i \overline{\mathbf{U}}_i \overline{e}_i\right) = \varepsilon_i \overline{\mathbf{T}}_i : \nabla \overline{\mathbf{U}}_i - \nabla \cdot \left[\varepsilon_i \left(\overline{\mathbf{q}}_i + \mathbf{q}_i^{\mathrm{Re}}\right)\right] + \varepsilon_i \overline{\rho}_i \overline{S}_i$$
$$+ \varepsilon_i D_i + E_i + e_{i,I}\Gamma_i \tag{2.5}$$

The final term in Eq. (2.3) represents mass transfer between phases. The last two terms in Eqs. (2.4) and (2.5) represent diffusive and convective transport between phases, respectively. These averaged quantities are conditioned with the gradient of the phase indicator function (Drew and Passman 1999),

$$\mathbf{F}_i = -\overline{\mathbf{T} \cdot \nabla X_i}, \tag{A22}$$

$$E_i = \overline{\mathbf{q} \cdot \nabla X_i}, \tag{A23}$$

$$\Gamma_i = \overline{\rho(\mathbf{U} - \mathbf{U}_I) \cdot \nabla X_i}, \tag{A24}$$

$$\mathbf{U}_{i,I}\Gamma_i = \overline{\rho\mathbf{U}(\mathbf{U} - \mathbf{U}_I) \cdot \nabla X_i}, \tag{A25}$$

$$e_{i,I}\Gamma_i = \overline{\rho e(\mathbf{U} - \mathbf{U}_I) \cdot \nabla X_i}, \tag{A26}$$

Considerable modeling effort has gone into finding expressions for the diffusive flux terms (Hao and Tao 2003b; Hill 1998; Rusche 2002; Drew and Passman 1999). The convective flux terms are often neglected (Hill 1998; Rusche 2002), but because phase change is important in the current study, proper handling of these terms is crucial.